HEALTH MANAGEMENT IN ASIAN AQUACULTURE

HEALTH MANAGEMENT IN ASIAN AQUACULTURE

Rohana P. Subasinghe
J. Richard Arthur
Mohamed Shariff

Reprinted by arrangement with the
Food and Agriculture Organization of the United Nations

by
Daya Publishing House
A Division of
Astral International Pvt Ltd
New Delhi - 110 002

"This book was originally published by the Food and Agriculture Organization of the United Nations (FAO) as Vaccine Manual, FAO Animal Production and Health Series No. 35".

First Indian Edition, 2005
Reprinted, 2025
ISBN: 978-81-70353-89-8

Published by : **Daya Publishing House**
A Division of
Astral International Pvt Ltd
4736/23, Ansari Road
Darya Ganj, New Delhi - 110002
Phone: 011-43549197
Email: info@astralint.com
Web: www.astralint.com

Printed at : **Chawla Offset Printers,** Delhi-110 052

PREPARATION OF THIS DOCUMENT

This document contains the technical papers presented at the Regional Expert Consultation on Aquaculture Health Management in Asia and the Pacific, jointly organized by the Fish Health Section of the Asian Fisheries Society (FHS/AFS) and the Fishery Resources Division (FIR) of the Fisheries Department of FAO, held at Universiti Pertanian Malaysia, Serdang, Selangor, Malaysia, 22-24 May 1995. The document also contains the recommendations made at the Expert Consultation. The full report of the Expert Consultation is published in the *FAO Fisheries Report* No. 529 (FAO, Rome, 1995, 24 p.).

Distribution:

FAO Fisheries Department
FAO Regional Fisheries and Aquaculture Officers
FAO field projects
Directorates of Fisheries
ᐨish Health Institutions of the Asia-Pacific
)ther interested parties
articipants.

Subasinghe, R.P.; Arthur, J.R.; Shariff, M. (eds.)
Health management in Asian aquaculture. Proceedings of the Regional Expert Consultation on Aquaculture Health Management in Asia and the Pacific. Serdang, Malaysia, 22-24 May 1995.
FAO Fisheries Technical Paper. No. 360. Rome, FAO. 1996. 142 p.

ABSTRACT

In 1994, world aquaculture production reached 25.5 million mt, valued at US$ 39.83 billion. Asia contributed 89.9% of this total, and has since continued to dominate global production. The drive to produce more fish and shellfish to meet the growing demand has lead many aquaculturists in Asia to intensify their operations. In many instances, the complex balance between the fish/shellfish and the environment is not well understood, the organism under culture subsequently becoming stressed and prone to infections. As we have already witnessed, disease has been and will continue to be a major constraint to the development of the aquaculture industry. Considering the FAO's priority on developing sustainable aquaculture, the large Asian contribution to global aquaculture production and the seemingly high losses of revenue due to diseases and health-related problems, FAO, in consultation with the Network of Aquaculture Centres in the Asia-Pacific (NACA), the Aquatic Animal Health Research Institute (AAHRI), the South East Asian Fisheries Development Centre (SEAFDEC) and the Universiti Pertanian Malaysia (UPM), and in collaboration with the Fish Health Section of the Asian Fisheries Society (FHS/AFS), organized a Regional Expert Consultation on Aquaculture Health Management in Asia and the Pacific, which was held at the Universiti Pertanian Malaysia in Serdang, Malaysia in May 1995. This document comprises the technical papers presented at the Consultation, and is a supplement to the report of the consultation, *FAO Fisheries Report* No. 529 (FAO, Rome, 1995. 24 p.)

(Key words: Asia, Pacific, Aquaculture, Fish disease, Health management, Quarantine)

CONTENTS

Page

CONCLUSIONS AND RECOMMENDATIONS OF THE REGIONAL EXPERT CONSULTATION ON AQUACULTURE HEALTH MANAGEMENT IN ASIA AND THE PACIFIC, HELD AT THE UNIVERSITI PERTANIAN MALAYSIA IN SERDANG, MALAYSIA IN MAY 1995 (*FAO Fisheries Report* No. 529, Rome, FAO. 1995. 24p.)

4.1 *General conclusions and recommendations*

4.1.1 *In light of the growing importance of aquaculture to global food supply and economic well-being and the highly significant contribution made by the Asian region to the world aquaculture production, the consultation* ***recognized*** *that it is imperative to ensure that the Asian aquaculture sector is developed in a sustainable manner.*

4.1.2. *Considering the past experiences with disease problems and the related economic losses to Asian aquaculture, the consultation* ***agreed*** *that diseases and related health problems are one of the most serious deterrents to the sustainable development of aquaculture in Asia and the Pacific.*

4.1.3. *It was* ***recognized*** *that not only health-related problems have a direct bearing on the aquatic environmental quality, but also that the future of the entire aquaculture industry will depend on the continuing availability of water and other resources with adequate quality.*

4.1.4. *The consultation* ***agreed*** *that there is an urgent need for an effective regional health management program and* ***suggested*** *three thrust areas for such a program; i) Research, diagnosis and information, ii) Training and extension, and iii) Quarantine and legislation.*

4.2. *Conclusions and recommendations on research, diagnosis and information*

Aquatic animal health research needs for a sustainable aquaculture industry in Asia have been recently reviewed at the Fish Health Management in Asia-Pacific Workshop organized by the ADB/NACA in Pusan, South Korea in October 1990 and at the workshop organized by the FHS/AFS and SIFR during the Second Symposium on Diseases in Asian Aquaculture held in Phuket, Thailand in October 1993. The consultation **endorsed** the specific recommendations made at the workshops and **emphasized** that these recommendations should be considered when formulating future aquatic animal health research programs for Asia. The consultation recognized the importance of regional cooperation in aquatic animal health research, diagnosis and information, and the following recommendations were made after carefully considering the past activities, present status, future directions, and the institutional and manpower capabilities in the region.

4.2.1. The Asian marine shrimp culture industry has expanded dramatically over the past decade. Although this growth is generated by market demand and short-term gains, and it represents only a small proportion of global aquaculture production, the contribution of this sector to the GNP of developing countries in the region is substantial. Recently, there have been a number of severe outbreaks of disease in Asian shrimp culture, mainly due to viral pathogens, which have caused serious economic losses to the regional shrimp industry. Considering the above:

The consultation ***reiterated*** *that viral diseases of shrimp were an extremely critical issue, more so than viral diseases of fish at present, and that immediate efforts must be made to develop research programs to combat viral disease problems in the Asian shrimp industry. It also* ***recognized*** *that medium- and long-term efforts should seek to develop the capacity for research on fish viruses, as such disease problems would likely occur in the future.*

4.2.2. In Asia, most countries do not appear to have adequate infrastructure for disease diagnosis and research on aquaculture health. Among the countries in the region, there are differences in prioritized species and systems for culture. The specific emphasis given by the governments in the region for establishing national aquatic health facilities appears to vary with their country's economic status. Therefore:

The consultation ***realized*** *that the primary responsibility for establishment and operation of national aquaculture health facilities, including diagnostic and research laboratories, remains with the national governments. Regional and international organizations should be encouraged to assist governments in strengthening such facilities.*

4.2.3. The disease diagnosis and research capabilities found in some countries in Asia and the Pacific are quite advanced. Trained diagnosticians, advanced laboratory facilities, and excellent aquaculture health centers are available and are functioning. Considering the fact that national-level capacity building in aquaculture health research and diagnosis is a long-term objective to achieve:

The consultation ***strongly emphasized*** *the importance of strengthening Asia's regional centers in aquaculture health management as a short-term activity. The participation of such centers in supporting capacity building in other countries of the region, through regional cooperation strategies, should be* ***strongly encouraged****. The consultation* ***agreed*** *that regional cooperative* *efforts towards strengthening aquaculture health centers could be further reinforced through increased international cooperation.*

4.2.4. Regional cooperation is encouraged to address aquaculture health problems in Asia. However, an effective regional networking mechanism with respect to aquaculture health management, which could bring effective regional cooperation, does not exist. With the expansion of regional aquaculture and the more frequent emergence of disease problems, a good networking mechanism could be very effective and would strengthen the regional aquaculture health management mechanism. Therefore:

The importance of regional networking was ***emphasized*** *as a cost-effective means of strengthening the national capacities for aquaculture health management. Increased efforts are required to promote the exchange of information, training, technical assistance and expertise to enhance such regional cooperative actions. These cooperative exchanges should include the more technically advanced countries of the region.*

4.2.5. In most Pacific countries, aquaculture is still in its infancy. However, the Pacific countries are interested in developing aquaculture and some have already embarked upon comprehensive programs. It is obvious that with the development of aquaculture in those countries, disease and health problems will emerge as major deterrents to the industry, if preventive and mitigating measures are not taken well in advance. Thus:

The consultation ***emphasized*** *that the Pacific countries could learn from experiences in the development of aquaculture in Asian countries.*

4.2.6. There are reports from the region about human zoonoses related to aquaculture. Some institutions are already addressing such problems, in particular the epidemiological aspects.

The consultation ***discussed*** *the possible merits of research on human health aspects of aquatic animal disease and* ***recognized*** *its importance to human well-being. However, it was noted that this subject was broader than aquaculture health management, and the consultation* ***recommended*** *that it be left to other initiatives.*

4.2.7. Genetic manipulation and the development of disease/pathogen resistant strains have been attempted with many fish and livestock species, but with limited success. The possible benefits of work on the genetic aspects of disease were discussed, particularly the possibility of incorporating disease resistance through selective breeding programs. Experience in the animal husbandry field suggested that such an approach was not promising. Thus,

*The consultation **recommended** that at this point in time research emphasis should not be given to the genetic aspects of disease.*

4.2.8. Effective dissemination of research findings, both regionally and internationally, is extremely vital to the advancement of scientific research. Many scientists in Asia appear to publish their research findings in local journals and reports, written in national languages. Although this grey literature has limited accessibility and use outside their country, it could provide vital information. Nevertheless, publishing research findings in peer-reviewed international journals will undoubtedly strengthen regional research.

*The consultation **endorsed** the importance of publishing fish health research findings in peer-reviewed journals and recommended that efforts should be made to **enhance** the availability and exchange of grey literature.*

4.2.9. The Asian Fish Health Reference Centre of the Fish Health Section of the Asian Fisheries Society has been established at the Universiti Pertanian Malaysia (UPM), where the bibliographic information pertaining to regional fish health topics is deposited. UPM has a fully fledged library with modern facilities and is willing to provide assistance to FHS/AFS to further strengthen its reference center.

*The consultation **strongly recommended** that participating countries supply aquaculture health literature to UPM, and that UPM should be encouraged to collate and disseminate such information to the countries of the region.*

4.2.10. Information technology in Asia, especially in many South East Asian countries, is sound, and many institutions are now capable of using it on a routine basis. These modern information technologies could be effectively utilized to help competent dissemination of fish health literature within and outside the region. Considering the availability of such advanced facilities in some countries in the region:

*The consultation **recommended** that the use of e-mail, bulletin boards and advanced information exchange technologies should be further explored as an effective means to exchange information on aquaculture health issues.*

4.3. ***Conclusions and recommendations on training and extension***

4.3.1. In most Asian universities, fish health is not taught within their fisheries, aquaculture, animal husbandry or veterinary curricula. There have been some attempts to incorporate fish health into university curricula, with little success. Considering the urgent need for trained manpower in the field of fish health to support sustainable development of regional aquaculture:

*The consultation **suggested** that universities should include aquatic animal health management components in curricula for fisheries, aquaculture and/or veterinary programs, but it was noted that very few appropriately skilled instructors are currently available. It was therefore **suggested** that, in the short term and until such expertise is more generally available, regional centers of excellence should provide the necessary training.*

4.3.2. There are many directories of experts in fisheries and aquaculture available in Asia and the Pacific. However, a concise directory on fish health scientists, providing their specialities and services, has yet to be produced. Such a directory would be of tremendous value in strengthening aquaculture health management in Asia and the Pacific. Considering the above:

*The consultation **recommended** that a directory of scientists in the region with expertise in aquaculture health should be produced. The listing should include specific information regarding each expert's skills.*

4.3.3. Training in fish health has been undertaken by a number of regional institutions. SEAFDEC (Philippines) and AAHRI (Thailand) are two institutions which conduct specific training programs for regional aquaculturists and extensionists on a regular basis. These institutions offer courses at different levels with different approaches addressing the specific requirements and needs of trainees and their countries. However, there has not been an attempt to compare and contrast the curricula of such courses offered by various institutions and to develop a regional training program. Considering the diversity of technical training courses offered in the region:

*The consultation **suggested** that attempts be made to combine the best features of the integrated health management approach (as in AAHRI programs) with the best features of the specialist discipline approach (as in SEAFDEC programs). Such integration could deliver more uniform training programs. It was **further recommended** that standardized curricula should be developed to assist national and regional centers in the development of training courses. It was **suggested** that training courses for trainers could be conducted in* *regional centers of excellence and/or more advanced countries such as Japan, UK, USA or Australia.*

4.3.4. Over the years, with the help of donor agencies, various Asian institutions have provided fish health training to a considerable number of regional aquaculturists and extensionists. However, it has been repeatedly highlighted that one of the major constraints to developing an effective health management strategy for regional countries is lack of trained personnel with practical experience. This raises the question as to the effectiveness of training programs. Considering the above:

*The consultation **recommended** that efforts should be made to evaluate the effectiveness of training courses and that such evaluation should be conducted by the training institution using independent evaluators.*

4.3.5. With respect to the type of training offered, field-based training for field personnel and laboratory-based training for laboratory workers was seen as important. However:

*The consultation also **recognized** that laboratory-based diagnosticians and researchers should be familiar with on-farm conditions and field-based extensionists should be familiar with diagnostics and therapy so that informed decisions on control and treatment can be made.*

4.3.6. Considering the severity of the recent disease outbreaks in Asia and their spread pattern:

*The consultation **recognized** the importance of epidemiology/epizootiology in providing solutions to aquaculture health problems, and noted the dearth of epidemiologists/epizootiologists currently working in the field. It was **recommended** that immediate efforts be made to rectify this situation.*

4.4. *Conclusions and recommendations on quarantine and legislation*

4.4.1. There is strong evidence that many of the disease outbreaks which have occurred recently in Asian aquaculture are linked to the introduction of new pathogens through trans-boundary movement of aquatic species. The distributions of many shrimp viral diseases and of epizootic ulcerative syndrome (EUS) appear to have direct relationships with the movement of species. As aquaculture grew over the past two decades, trans-boundary movement of aquatic organisms became intense, creating the importance and urgency of establishing quarantine systems, protocols and guidelines for the movement of aquatic species. Together with efforts to implement stringent consumer protection strategies and to ensure sustainable development of global aquaculture, guidelines, protocols and codes of practice for aquatic animal health quarantine and movement have been developed by various agencies and institutions. After careful consideration of the recent global and regional developments with respect to aquatic animal health and quarantine:

*The consultation **strongly emphasized** the importance of documenting economic justifications and cost-benefit analyses for introduction of legislation and associated quarantine systems. Policy makers are only likely to be convinced by strong economic arguments. Marketing and trading issues are also important concerns in relation to the overall transfer of aquatic animals and pathogens from one place to another.*

4.4.2. In order to develop practical and realistic guidelines and legislation on quarantine and health, it is imperative to investigate carefully the regional fish health status, including current trends in the movement of species, occurrence and pattern of spread of diseases, causative agents and their origin, and existing guidelines and legislation. It is perceived that the required literature is available in Asia and that this information needs to be compiled into an effective format. Thus:

*The consultation **strongly recommended** that a literature search be conducted, as soon as possible, to develop a preliminary list of pathogens of aquatic animals in the countries of the region. The list would help in the assessment of quarantine and aquatic animal health legislation. However, it was **recognized** that further research on the epizootiology and epidemiology of aquatic animal diseases will be required to develop a comprehensive list.*

4.4.3. In an effort to facilitate international trade in aquatic animals and animal products, the Office Internationale des Epizooties (OIE) will soon publish the International Aquatic Animal Health Code and Diagnostic Manual, the prototype of which is now being reviewed. The Code and Manual attempts to achieve this aim by providing detailed definitions of minimum health guarantees, based on standard methods for laboratory examinations, to be required for trading partners in order to avoid the risk of spreading aquatic animal diseases.

*The consultation **discussed** the OIE Aquatic Animal Health Code and Diagnostic Manual. It was **emphasized** that the Code and Manual are based on temperate species and aquaculture systems in developed countries, and that care should be taken in applying these guidelines to the Asian region. The consultation **stressed** the importance of evaluation of the OIE Code and*

Manual in relation to the situation in the Asian region, and their further development to address the species and systems of Asia.

4.4.4. Over the years, certain South East Asian countries such as Indonesia and Malaysia have attempted to introduce aquatic animal quarantine procedures and to adopt appropriate legislation. Many other countries in the region are now in the process of developing quarantine guidelines with the view to making them mandatory for movement of aquatic animals and animal products both locally and internationally. It was noted that these attempts have been only partly successful and certain countries find it difficult to implement such guidelines. Considering the regional status of aquatic animal quarantine:

*The consultation **recommended** that an assessment of `case studies' of successful and unsuccessful attempts at fish quarantine be made part of the process of developing guidelines for governments on quarantine/aquatic animal health legislation. The importance of regional exchange of information on the subject was **recognized**.*

4.4.5. In order to produce practical quarantine guidelines, it is imperative that standardized diagnostic methods which can be adopted by the countries of the region be developed. Although such standardized procedures are in place elsewhere, their applicability to the Asian region is questionable. Therefore:

*The consultation **recognized** that the development of high-tech methods for assessing the disease status in Asia would be useful. For example, there are techniques available in Australia for testing for carriers of Aeromonas salmonicida, which might be usefully applied. Polymerase Chain Reaction (PCR) probes and other techniques, including immunological and genetic methods of detecting small numbers of organisms, might also be used in Asia.*

4.4.6. The issue surrounding the trade in aquatic animals and animal products and their disease status will become increasingly linked and important. Certification procedures for fish health are being developed by various agencies and bodies, such as the European Union (EU) and the World Trade Organization (WTO). Considering the recent development of such certification procedures:

*The consultation **recommended** that, in the region, efforts should be made to ensure that any future certification procedures be workable and relevant to the species and the systems of the countries in the region.*

4.4.7. Considering the urgency of developing an effective regional fish health management strategy for Asia and the Pacific:

*The consultation **emphasized** the need for a broad assessment of quarantine and fish health legislation, with a view to developing guidelines for assisting governments in policy development. The assessment would involve an analysis of regional experiences and an evaluation of the applicability of existing international codes, including those of OIE, the European Inland Fisheries Advisory Commission (EIFAC), and the International Council for the Exploration of the Sea (ICES). Such an assessment should be carried out through a regional cooperative effort, with the assistance and cooperation of concerned international agencies and bodies, which might include OIE, EIFAC, NACA and FAO. The assessment should cover ornamental fish as well as fish and shellfish produced for human consumption. The **importance** of involving the industry/private sector in such activities was underlined.*

4.4.8. Since all countries in the region involved in aquaculture production are concerned with the sustainable development of the industry through prevention and control of disease:

*The consultation **agreed** that national aquaculture policies should incorporate relevant aquatic animal health issues. These issues were detailed in the earlier recommendations made by the ADB/NACA Workshop in 1990.*

Better Health Management in the Asia-Pacific through Systems Management

Michael J. Phillips

Network of Aquaculture Centres in Asia-Pacific
P.O. Box 1040, Kasetsart Post Office
Bangkok, Thailand

Phillips, M.J. 1996. Better health management in the Asia-Pacific through systems management. *In* Health Management in Asian Aquaculture. Proceedings of the Regional Expert Consultation on Aquaculture Health Management in Asia and the Pacific. R.P. Subasinghe, J.R. Arthur & M. Shariff (eds.), p 1-10. FAO Fisheries Technical Paper No. 360, Rome, FAO. 142 p.

Abstract

This paper provides an overview of systems approaches to aquatic health management. It briefly reviews issues relating to aquatic animal health and the environment; environmental risk factors in aquaculture contributing to aquatic animal health problems; advantages of systems approaches to the prevention of aquatic animal diseases; and the implications of systems approaches for research, training and extension.

INTRODUCTION

This paper discusses the potential for better heath management of cultured aquatic animals through aquaculture systems management. In farming systems research, the interactions between the different types of components of the sub-systems must be identified and understood if the system as a whole is to be properly understood and managed (Deomampo, 1995). In the context of aquatic animal health, the systems management approach implies an understanding of the ways in which the different components of the aquaculture system interact with one another and the management of these components in a way which provides conditions: (a) optimal for the cultured animals (or plants); and (b) least favorable for the occurrence of disease. The approach relies heavily on environmental management but also includes "non-environmental" factors, as noted in the following list of system "components":

- soil, water quality, climate, pond location and other factors;
- biological components, including the cultured animal (or plant), stress levels, nutrition and feeding practices, genetic factors, the presence or absence of pathogens and occurrence of plankton bloom;
- human components, such as the management experience of the farmer, access to capital and other resources, traditions and socio-cultural aspects;

- legal/institutional components, which influence the allocation of resources and institutional support, as well as laws *e.g.*, quarantine, trade laws, and access to information, training and appropriate expert assistance; and
- economic components, which determine the supply and demand situation for aquaculture products, and influence farm management decisions which have an impact on aquatic animal health *e.g.*, the degree of attempted intensification.

NEED FOR A SYSTEMS APPROACH

The introduction of many factors into the overall aquatic animal health management picture may appear to some to complicate the management situation greatly. However, there are reasons for taking a broader approach which considers the management of the system, rather than individual components:

- aquatic animal disease is the end result of a series of linked events, therefore, treatment of disease goes beyond consideration of the pathogen. The heavy focus on the pathogen in research, diagnosis and extension can lead to inappropriate and ineffective "cures" which may temporarily remove the symptom, but not the cause (*e.g.*, such an approach can lead to reliance on inappropriate chemicals and other "quick fix" solutions);
- current approaches have so far had limited success in the prevention or cure of aquatic disease. Considerable difficulties remain in dealing with some existing problems, such as the recent economically serious diseases which have emerged in the shrimp industry; and
- large-scale outbreaks of aquatic animal disease, particularly in shrimp, demonstrate the importance of the linkage with other farms *i.e.*, the problems easily spread beyond the management of individual farms, requiring broader ecosystem management approaches to control farm level environmental deterioration and to take preventative measures to avoid introduction of pathogens.

AQUATIC ANIMAL HEALTH, STRESS AND THE ENVIRONMENT

- The environment in which an animal (or plant) is cultured plays a critical role in the degree to which that animal is susceptible to pathogens and the occurrence of clinical disease (AAHRI, 1994; Flegel and Sriurairatana, 1993), and an understanding of the relationship between host, pathogen and environment is important for understanding the cause, prevention and treatment of most aquatic animal diseases. In general, diseases affecting aquatic animals may be grouped into:

- disease resulting from poor environmental conditions leading to direct effects (*e.g.*, low dissolved oxygen, toxins and red tides);
- disease resulting from stress leading to infection by opportunistic pathogens (*e.g.*, vibriosis - an economically important disease of shrimp);
- pathogens causing disease only when animals are stressed (*e.g.*, MBV in shrimp); and
- primary pathogens causing disease without environmental stress. These are comparatively rare, although some recently reported shrimp viral infections such as yellow head baculovirus may fall into this category (even then, environmental management may be required to control entry of such pathogens to the culture system).

The relationships between the environment and the occurrence of aquatic animal diseases are poorly understood, particularly the relations between stress and disease occurrence. For example, it is recognized that shrimp are "stressed" by poor environmental conditions, but the interaction between disease occurrence and stress in shrimp is unclear (AAHRI, 1995). However, enough is known of the general conditions under which healthy aquatic animals exist that the lack of research data is not a major constraint to improved health management using the systems approach. For example, water quality, pond bottom conditions, feed and plankton bloom management are widely recognized as critical for disease prevention in shrimp culture, particularly for commonly occurring opportunistic pathogens (Flegel and Sriurairatana,1993).

ENVIRONMENTAL IMPACTS AND RISKS

An important and growing concern (*e.g.*, see FAO/NACA, 1995) is the deteriorating environmental condition of some aquaculture environments, placing an additional stress on cultured organisms. The environmental impacts affecting aquaculture systems include:

- those inherent in the site (*e.g.*, water and soils);
- environmental impacts ***on*** aquaculture *i.e.*, those caused by non-aquaculturists, such as through industrial, domestic and agricultural pollution;
- environmental impacts ***of*** aquaculture *i.e.*, through discharge of pond effluent and habitat changes caused by pond construction; and
- impacts of aquaculture ***on*** aquaculture *i.e.*, through self-pollution of water supplies through discharge of pond effluent.

An important concern is the impact of aquaculture on the environment which has been linked to outbreaks of serious disease (see the comprehensive reviews in FAO/NACA, 1995). For instance, self-pollution caused by effluent from shrimp ponds is thought to have contributed to serious disease outbreaks in Thailand, China and India, and it is probably no coincidence that the first serious outbreak of shrimp disease in India was centered on shrimp farms around Kandleru Creek, a poorly flushed tidal creek suffering from self-pollution caused by overcrowding of ponds (India country report in FAO/NACA, 1995). In many cases, however (such as the upper Gulf of Thailand), the impacts of aquaculture effluent are difficult to separate from environmental deterioration caused by non-aquaculture sources. Such problems further highlight the point that aquaculture farms (and their management) cannot be considered in isolation from the surrounding environment.

Existing experiences with such environmental problems and disease outbreaks allow certain conditions to be identified when aquaculture farms have increased risks of environmental problems giving rise to disease. The following need to be considered in assessing environmental risk:

1. Aquaculture system and management

The type of culture system and its management is obviously important, and the following factors affect risk:

Intensification of farming, through increased stocking densities and feed inputs, leading to:

- increased risk of water and soil quality deterioration; and
- increased need for on-farm environmental management (and management expertise).

Open *versus* closed systems:

- open systems such as cage farms and flow-through pond/raceways systems are exposed to greater risk of external environmental impacts; and
- closed systems offer greater opportunities for environmental control and management.

Aquaculture system diversity:

- monoculture systems have reduced diversity and stability, increasing the likelihood of water quality and soil problems;

Degree of reliance on external inputs (*e.g.*, feed and seed):

- increased reliance on external inputs is related to intensification, and also increased risk of importing pathogens onto aquaculture farms;

Farm management expertise (*e.g.*, new entrants to the industry with limited experience are likely to suffer greater problems than experienced farms, as is particularly noticeable in shrimp farming).

2. Intensification of natural resource use

The intensity with which aquaculture uses resources, and how aquaculture competes with others for those resources, is important with:

- increasing resource use pressure leading to impacts on aquaculture from increased number of competing users (*e.g.*, agricultural, industrial and domestic users), leading to increased risk of pollution problems. The intense development activities in many areas of Asia, and particularly in coastal areas, are and will continue to lead to environmental deterioration impacting on aquaculture;
- increasingly severe impacts of aquaculture when the resources utilized for aquaculture exceed the capacity of the environment to supply these resources *e.g.*, when self-pollution occurs due to excessive development of cage farms or shrimp ponds in areas with limited water flushing capacity.

3. International Trade

An over-riding influence, and beyond the control or management of individual farmers, is that related to international trade and the liberalization of trading policies within and between countries and regions. Such issues are important in terms of:

- Transboundary movement of aquatic animals:

  The increasing movement of live aquatic animals (and plants) for aquaculture and the aquarium trade increases the risks of the rapid spread of pathogens and introductions of exotic species. Interestingly, such risks are increasing at a time of increased concern to maintain biological diversity and heightened awareness of the risks of uncontrolled movement of aquatic animals; and

- Exchange of technology:

  Rapid spread and introduction of technologies without consideration of environmental impacts increase the risk of environmental problems.

SYSTEMS APPROACHES

The emphasis of a systems approach should be on prevention rather than cure, which is likely to be the most cost-effective, involving both on-farm management and the management of the broader environment, and involving both aquaculturists and governments. Whilst individual aquaculturists are responsible for farm management, the government inputs are essential for regulation of the resource use, particularly land and water, and for helping to provide legal and institutional arrangements which minimize resource use conflicts and environmental impacts (of and on aquaculture).

On-farm environmental management options include:

- Location/site selection concerns
- Operational concerns:
  - water management, including plankton bloom management;
  - soil/sediments;
  - feed (quality and quantity);
  - seed (disease free/stocking density);
  - chemical use; and
  - monitoring and record keeping.

The emphasis on farm management approaches is on: (a) maintaining a suitable environment for the cultured animal; and (b) reducing risks from introduction of infectious agents.

Off-farm management essentially involves the management of ecosystems which support aquaculture, and considerations of environmental impacts, including:

- Reduction or management of environmental impacts within the ecological limits:
  - effluent management and control;
  - controlling risks associated with introduction of pathogens/exotics species.

- Control of environmental impacts on aquaculture:
  - pollution control strategies;
  - reducing user conflicts through fair and equitable allocation of resources; and
  - integration of aquaculture into broader watershed/coastal management strategies.

ADVANTAGES OF SYSTEMS APPROACH FOR ON-FARM AQUATIC HEALTH MANAGEMENT

There are many advantages in further promoting a systems approach towards prevention and cure of aquatic animal diseases, including:

- a more effective solving of aquatic animal disease problems, because the systems approach considers management of the factors leading to disease (the "cause" rather than the "symptom");
- systems approaches can promote farmer understanding of aquaculture systems and the environment, increasing the capacity to solve future problems, and reducing the reliance on "quick fixes";
- systems approaches can lead to increased self-reliance by farmers and less reliance on external support from *e.g.*, scientists, veterinarians, and extension officers. Given the lack of sufficient manpower resources, an important consideration for aquatic animal health practitioners should be to reduce, not increase, reliance of farmers on "professional" inputs for disease control and farm management;
- solutions developed through a systems approach are more likely to be workable and appropriate, and contribute more to sustainable development;
- systems approaches can be more cost-effective, as they do not rely on costly inputs (such as chemicals) and can lead to lasting improvements.

IMPLICATIONS FOR TRAINING AND EXTENSION

The adoption of a systems approach raises some important issues also for training and extension of information to farmers:

- training of aquatic animal health practitioners could give stronger emphasis to the systems management approach to diagnosis and control of disease; and
- systems approaches can be used to better define appropriate information requirements and the most effective channels for providing information to farmers.

The adoption of a systems (or more "holistic") approach in aquatic animal health training, based largely on managing the environment in a way which ensures it is suitable for the animal and not for the pathogen, is increasingly being recognized as an important way forward in disease control (see also the excellent manual by AAHRI - which exemplifies this approach - AAHRI (1995)). As mentioned above, this approach also puts disease control within the hands of the farmer, encouraging self-reliance, sustainability and development of farm level and appropriate solutions.

Training which provides only for the identification of the pathogen (often involving complex procedures) without providing the farmer with an appropriate management strategy should be discouraged. The increasingly popular diagnostic "kits" for shrimp culture, for example, should be linked to some management advice, rather than be used solely for pathogen identification.

A systems approach can be taken in the dissemination of appropriate information to farmers. An understanding of the farming system will allow appropriate information to be identified. An understanding of institutional linkages and where farmers obtain information from can be used to identify optimal strategies for dissemination of information.

One problem is the difficulty of making contact with large numbers of small-scale farmers, with existing manpower and resources. Whilst improvements in government services might help, efforts are warranted to explore other means of extending "messages" to farmers through more "remote" methodologies. In Thailand, approaches include radio, television, posters and other materials for shrimp farmers. Studies have shown chemical and feed salesmen are the most frequent visitors to small-scale shrimp farms in Thailand - and the source of most information on shrimp farm management (Office of the Environmental Policy and Planning, 1994). Information linkages with feed salesmen or other commercial outlets could perhaps be explored, whilst trying to avoid some of the obvious conflicts of interest which might arise in such an approach (FAO/NACA, 1995).

IMPLICATIONS FOR AQUATIC ANIMAL HEALTH RESEARCH

The adoption of systems approaches to aquatic animal health management has implications for identification of research needs and for the implementation of research projects. It is common for research scientists to set their own priorities, indicating widespread misuse of the term "demand led." However, the importance of research being "led" by the demand of the end user (not the scientist looking for funding for research) is increasingly being recognized. This applies equally to aquatic health management research.

Aquaculture researchers are increasingly realizing the benefits of identifying research needs based on systems approaches and through talking to farmers (see Pullin, 1993). Further considerations in applying a systems approach to research include:

- more favorable cost-benefit from systems research, and research results likely to be more adoptable by farmers;

- timely and appropriate dissemination of results in a form suitable for "end users" taking into account language and educational background; and
- researchers need to consider the costs which may be borne by governments and/or farmers in the application of research results.

The systems approach goes further beyond the identification and implementation of research. The adoption of systems approaches to research requires cooperation involving multi-disciplinary teams. As the capacity of individual institutes may be limited, such an approach may require cooperation between institutes within countries, and sometimes cooperation between institutes in different countries. There are already some examples of fruitful cooperation among institutes in different parts of the world which have brought strong benefits in problem solving (*e.g.*, that adopted in tackling epizootic ulcerative syndrome, EUS).

Some more specific areas where systems research might be undertaken include:

- environment-animal interactions, and optimal environmental control strategies for low-stress, low-disease, aquaculture systems; and
- assessment and management of environmental impacts and the development of low impact systems (which retain/enhance farm profitability and sustainability).

SUMMARY

This paper discusses the application of a "holistic" systems approach to aquatic animal (and plant) health management. This approach implies an understanding of the links which lead to the occurrence of outbreaks of disease, and management of the aquaculture system in a way which reduces the stress and the risk of occurrence of disease (ADB/NACA, 1991). This approach implies on-farm and off-farm management, within ecological limits. Such approaches, particularly if more widely adopted at the on-farm level, are likely to lead to sustainable solutions to aquatic disease problems which can be adopted by farmers, and less reliance on the use of chemicals which largely treat the symptom of the problem and not the cause. In addition, research, training programs, extension and information exchange can be more effective and responsive to farmers' needs if based on systems approaches to the understanding and management of aquaculture farms.

REFERENCES

AAHRI, 1995. Health management in shrimp ponds. 2nd Edition. Aquatic Animal Health Research Institute, Bangkok, Thailand.

ADB/NACA, 1991. Fish Health Management in Asia-Pacific. Report on a Regional Study and Workshop on Fish Disease and Fish Health Management. ADB Agricult. Dep. Rep. Ser. No. 1. Network of Aquaculture Centres in Asia-Pacific, Bangkok, Thailand.

Deomampo, N.R., 1995. Farming systems, marketing and trade for sustainable aquaculture. *In* Report of the ADB/NACA Regional Study and Workshop on Aquaculture Sustainability and Environment. Network of Aquaculture Centres in Asia-Pacific (NACA), Bangkok, Thailand.

FAO/NACA, 1995. Regional Study and Workshop on the Environmental Assessment and Management of Aquaculture Development. Food and Agricultural Organization of the United Nations and Network of Aquaculture Centres in Asia-Pacific. Bangkok, Thailand.

Flegel, T.W., and S. Sriurairatana, 1993. Shrimp health management: an environmental approach, p. 1-48. *In* R.P. Subasinghe and M. Shariff (eds.) Diseases in Aquaculture: the Current Issues. Malaysian Fish. Soc. Publ. No. 8.

Office of Environmental Policy and Planning, 1994. The Environmental Management of Coastal Aquaculture. An Assessment of Shrimp Culture in Southern Thailand. Prepared for the Office of Environmental Policy and Planning by the Network of Aquaculture Centres in Asia-Pacific. Network of Aquaculture Centres in Asia-Pacific, Bangkok, Thailand.

Pullin, R.S.V., 1993. An overview of environmental issues in developing-country aquaculture, p. 1-19. *In* R.S.V. Pullin, H. Rosenthal and J.L. Maclean (eds.) Environment and Aquaculture in Developing Countries. ICLARM Conf. Proc. 31, 359 p.

Fish and Shellfish Quarantine: the Reality for Asia-Pacific

J. Richard Arthur

Department of Fisheries and Oceans, Maurice Lamontagne Institute
P.0. Box 1000, Mont-Joli, Quebec, Canada G5H 3Z4

Arthur, J.R. 1996. Fish and shellfish quarantine: the reality for Asia-Pacific. *In* Health Management in Asian Aquaculture. Proceedings of the Regional Expert Consultation on Aquaculture Health Management in Asia and the Pacific. R.P. Subasinghe, J.R. Arthur & M. Shariff (eds.), p. 11-28. FAO Fisheries Technical Paper No. 360, Rome, FAO. 142 p.

Abstract

This paper reviews progress made in the Asia-Pacific Region towards establishment of quarantine systems for aquatic organisms. Quarantine programs form part of a first line of defense against possible adverse effects resulting from the introduction or transfer of exotic fish and shellfish. As such, they must be developed within the context of larger national and international plans addressing this problem. "Codes of Practice" for the introduction and transfer of aquatic organisms which have been developed by international organizations provide a starting point for designing national fish health legislation and international agreements to prevent the spread of disease. To succeed, such efforts must be accompanied by the development of regionally agreed-upon lists of certifiable pathogens, the standardization of diagnostics techniques and the production of health certificates of unambiguous meaning. Strong commitment by national governments and the cooperation of importers/exporters are considered key elements in the success of these programs. Successful disease prevention will also be directly related to the ability of countries to reduce their dependence on imported broodstock and fry for the aquaculture industry, and shipments for the ornamental fish trade, particularly those involving wild-caught fishes.

INTRODUCTION

Quarantine can be defined as the holding or rearing of animals under conditions which prevent their escape or the escape of organisms and potential disease agents infecting or associated with them into the natural environment. Quarantine programs for aquatic organisms typically involve protocols for inspection i.e., the examination of animals for disease agents, and certification, the issuing of a certificate stating that a particular lot of animals or a production facility has been inspected and is free from infection by a particular pathogen or pathogens.

Properly designed quarantine programs may be effective at several levels. Internationally and nationally, they are aimed at preventing the spread of exotic species or strains of parasites, bacteria and viruses into countries where they do not occur, while at the sub-national level (state, prefecture, etc.) they help to reduce within-country dissemination of pathogens, be they native or introduced. Finally, at the local level, quarantine of fry and broodstock originating from an outside source helps to protect individual government hatcheries and private fish farms from potentially devastating losses caused by disease, and the conscientious home aquarist from less serious financial setbacks.

At the international and national levels, which are those to which this paper is directed, quarantine programs form an integral part of much broader strategies aimed to protect the natural environment and native faunas from the deleterious impacts of exotic species. They serve the aquaculture industries by protecting them from diseases which the industries themselves might introduce when importing fry or broodstock. They also protect human health by preventing the entry and possible establishment of exotic parasites infective to man which are transmitted by aquatic organisms.

INTRODUCTIONS - THE BROAD PICTURE

Ill-considered or accidental introductions of exotic species or strains of aquatic organisms can adversely affect local faunas in a variety of ways. These include genetic pollution, disease introduction and ecological impacts, such as predation, competition and environmental modification (see de Kinkelin *et al*., 1985; Sindermann, 1986, 1992; De Silva, 1989; Stewart, 1991; Porter, 1992; Williams and Sindermann, 1992; Carlton, 1993 a,b).

Exotic species are often intentionally introduced, in the hope that they will improve fish production or control nuisance species such as aquatic weeds and mosquitoes. They are also imported for aquaculture purposes and to increase commercial and sport fishing opportunities. Such intentional introductions are done at all levels, ranging from international agencies and national governments to universities, local sportsman's groups and individuals. Accidental introductions also frequently occur, through escapes from culture facilities or backyard ponds (e.g., tropical aquarium fishes) and by transport on boats and in their ballast waters. Indeed, the movement of aquatic organisms in ships' ballast water has recently been recognized as a major cause of undesirable introductions (see Carlton, 1993b), including that of the notorious zebra mussel and the European ruffe into the Great Lakes system in North America.

All wild and most cultured fish and shellfish populations carry parasites, bacteria, viruses and other potential pathogens. When they are moved to new areas, the inevitable result is that disease agents are also transferred, a fact amply documented in the many reports detailing the international spread of fish and shellfish parasites and pathogens as a result of human activities, and their harmful and occasionally devastating effects on native populations (see, for example, Hoffman, 1970; Bauer and Hoffman, 1976; Williams and Sindermann, 1992).

Examples of pathogens which have caused important losses of fish and shellfish in the Asia-Pacific Region and which are believed to be introduced include the copepod *Lernaea cyprinacea* and myxosporeans of the genus *Myxobolus*, which have caused problems in Indonesia, Epizootic Ulcerative Syndrome (EUS), which has spread throughout much of South and Southeast Asia, and several viral diseases of penaeid shrimp which are a continuing problem across the region and, indeed, to prawn growers around the world (see Djajadiredja *et al.*, 1983; Lightner, 1985, 1990; Arthur and Shariff, 1991; Lilley *et al.*, 1992).

Economic losses caused by pathogens are difficult to quantify, but are increasingly recognized to be substantial (see Shariff, 1995). For example, Tonguthai (1985) estimated that EUS-caused losses amounted to US$ 8.7 million in Thailand during 1982-1983 alone. More recently, China, which in 1991 was the world's largest producer of farmed shrimp with 200 000 mt, has seen its production reduced by 75% due to a new and possibly exotic disease, China baculovirus (Chamberlain, 1994). It has been estimated that total losses caused by disease in 15 Asian countries in 1990 amounted to at least US$ 1.36 billion, or approximately 6% of the total value of all aquaculture output (ADB/NACA, 1991). Many of the disease agents involved in these losses are known or believed to have been introduced by human activities.

INTERNATIONAL *CODES OF PRACTICE*

Measures taken by national governments to prevent the introducuion of exotic fish and shellfish diseases must be developed within the context of larger national and international policies dealing with the introduction and transfer of plants and animals. The Office International des Épizooties (OIE), an international veterinary organization with 139 member countries, has recently developed revised recommendations and protocols for the prevention of the international spread of diseases of aquatic organisms which are given in its new *International Aquatic Animal Health Code* (S.N. Chen, National Taiwan University, Taipei, *pers. comm.*). Recommendations for policies dealing with the introduction of aquatic species and guidelines for their implementation, including methods to minimize the possibility of disease transfers, have also been developed by the International Council for the Exploration of the Sea (ICES) for marine introductions (see Anon., 1984; Carlton, 1993c) and the European Inland Fisheries Advisory Commission of the FAO (EIFAC) for transfers of both marine and freshwater organisms (Turner, 1988). More regionally oriented guidelines are provided by the Great Lakes Fish Disease Control Committee of the Great Lakes Fishery Commission (Meyer *et al.*, 1983) and the North American Commission of the North Atlantic Salmon Conservation Organization (Porter, 1992), among others.

An example of such recommendations is the Revised 1990 *Code of Practice to Reduce the Risks of Adverse Effects Arising from the Introduction and Transfers of Marine*

Species which has been developed by the ICES *Working Group on Introductions and Transfers of Marine Organisms* (see Carlton, 1993c). The ICES *Code of Practice* is divided into five major parts:

1. A recommended procedure for all species prior to reaching a decision regarding new introductions;

2. Recommended action if the decision is taken to proceed with the introduction;

3. A suggestion that regulatory agencies use the strongest possible measures to prevent unauthorized introductions;

4. A recommended procedure for introduced or transferred species which are part of current commercial practice; and

5. A note recognizing that countries will have different attitudes toward the selection of the place of inspection and control of the consignment.

In general, these Codes of Practice and their supporting documents attempt to establish specific diagnostic techniques, define and make known the sanitary regulations of individual countries, develop health certificates whose meanings are unambiguous, and prohibit the international transfer of fish which are not accompanied by these certificates (see de Kinkelin *et al.*, 1985).

The approach relating to disease control in cases of exotic fish introductions as recommended in the EIFAC *Code of Practice* can be summarized as follows:

1. Examine each proposed transfer for the possibility of introducing pathogenic organisms and parasites;

2. Establish a brood stock in the importing country by transfer of eggs to an approved quarantine facility where they will be examined regularly for pathogens;

3. If no pathogens become evident, transplant first generation progeny, but not the original import, to culture sites or the natural environment;

4. Continue disease studies on these transplanted individuals.

An important additional step recommended by the ICES *Code of Practice* is to eliminate the need for further transfers by using F_1 individuals to establish a local brood stock.

Such stringent measures are justified to protect existing aquaculture development and avoid harmful effects to indigenous species, both those important to commercial and recreational fisheries and those not currently exploited. The latter may be particularly vulnerable to introduced pathogens, as not having been previously exposed to infection, they often lack natural resistance.

The procedures for fish quarantine recommended in the Code were essentially formulated among developed nations, and have not been widely implemented. Few developing nations have the necessary resources to undertake such stringent measures. Implementation of the Code has sometimes been considered onerous. However, this finding, in itself, indicates that the introduction being contemplated should probably not take place.

PROBLEMS IN IMPLEMENTATION

The implementation of effective quarantine programs by developing countries is difficult due to the high financial requirements and the need for highly trained technical expertise. Some of the requirements are:

1. Highly trained personnel (fish disease diagnosticians, backed by services in parasitology, virology, mycology, bacteriology and water chemistry);
2. Adequate diagnostic and quarantine facilities;
3. Detailed knowledge of potential pathogens, including aspects of their biologies, host specificities, pathologies and geographic distributions;
4. Detailed knowledge of the indigenous fauna;
5. Effective legislation;
6. Adequate enforcement;
7. Enlightened and cooperative aquaculture and aquarium fish industries;
8. Existence of a communications network of experts.

In developing countries funds are scarce. The cost/benefit ratio must be high enough to convince government planners of the need for quarantine and inspection services, yet detailed economic analyses of the losses caused by introduced pathogens are few.

Knowledge on fish diseases is restricted to a very few species, less than 2% of the total number known to science (Kinne, 1984). In the Asia-Pacific Region, as elsewhere, little or no information exists for the vast majority of aquatic organisms. Only when basic work on the identities, biologies and distributions of pathogens occurring in the region has been accomplished can meaningful lists of certifiable pathogens be compiled and reliable diagnostic techniques developed.

Aquaculturists and aquarium fish importers all too often regard quarantine and inspection programs as unnecessary evils inflicted upon them by government bureaucracies out to hinder their legitimate business activities. More efforts need to be made to show these groups that such programs will be to their benefit, helping to protect their industries and make them more profitable by reducing losses due to disease. These clients should thus be consulted early on during the development of quarantine and inspection programs, so that they will recognize the need for these programs and support their development and implementation.

PRIORITIES AND NEEDS

Various attempts have been made by regional and international fish health experts to identify priorities for and constraints to the development of quarantine and certification programs for the region and/or to identify needs related to the general development of fish health expertise (see Davy and Graham, 1979; Davy and Chouinard, 1983; Arthur 1987, 1995; Arthur and Shariff, 1991; ADB/NACA, 1991; Langdon *et al.*, 1992). The following list summarizes the recommendations related to legislation and the control of fish diseases made by scientists attending the Asian Development Bank/Network of Aquaculture Centres in Asia (ADB/NACA) *Regional Study and Workshop on Fish Disease and Fish Health Management* held in Bangkok in 1991 (Wootten, 1991). These experts agreed that countries of the region should:

1. Prepare legislation to prevent the translocation of serious fish diseases both within and outside the region;

2. Develop the capability of testing exports of fish to an agreed upon regional standard;

3. Develop quarantine systems where imports of fish may be tested to regional standards;

4. Establish a standardized system of disease testing, including a common format of health certificate;

5. Compile a regional handbook of diagnostic methods;

6. Quarantine and test for disease, introductions of new species in accordance with the ICES *Code of Practice*;

7. Establish a working group of regional and international experts to deal with the above recommendations.

PROGRESS TOWARDS QUARANTINE WITHIN THE ASIA-PACIFIC REGION

This paper focuses primarily on the tropical/sub-tropical areas of the Asia-Pacific and thus considers Australia and Papua New Guinea, and South, Southeast and East Asia. The status of quarantine and certification programs in this region has been summarized on a number of occasions (Davy and Graham, 1979; Davy and Chouinard, 1983; Arthur, 1987, 1995; Arthur and Shariff, 1991; ADB/NACA, 1991). Information presented below for the various countries is summarized mainly from these papers.

Australia and Papua New Guinea

Australia is the only country among those considered herein which has a fully operational quarantine system for imported fishes. The existing system, which was established in 1984, prohibits the entry of live fish unless the species to be imported is listed by the Australian Nature Conservation Agency (ANCA) (formerly the Australian National Parks and Wildlife Service) as being among those aquarium fishes which may be freely imported, or which can be imported under special permit for scientific purposes (see Schmidt and Love, 1985; Humphrey, 1986; O'Connor, 1990; Lehane, 1992). Australia requires that freshwater ornamental fish be accompanied by a declaration of health issued by the exporting country. The premises of the exporter must be inspected regularly, and the exporter certified as competent, by government fish health inspectors of the exporting country. A system of registration of exporters and importers of freshwater ornamental fish has been established and imported freshwater species are held in licensed and inspected quarantine facilities which are provided at the expense of the importer. Fish are held for a minimum of 14 days, during which time they are examined regularly for clinical signs of disease by officers of the Australian Quarantine and Inspection Service (AQIS), the costs of examination being borne by the importer. The AQIS is empowered to require that laboratory examinations be conducted, and does so from time to time. In cases where infected fish are found, the AQIS may treat and reinspect them at the importer's expense, or destroy them without compensation.

Importers and exporters of marine aquarium fishes are not registered (P. Beers, AQIS, Canberra, *pers. comm.*). Marine species are not quarantined upon arrival, but are inspected to determine their identity, to verify that they are among the species approved for importation, to ensure that they exhibit no clinical signs of disease and to confirm that no other organisms are present in the shipment (Schmidt and Love, 1985; Lehane, 1992). Various regulations also apply to the importation of non-living material derived from or associated with aquatic animals.

The current Australian system, like that in place, in the process of establishment, or being envisioned by various countries of the Asia-Pacific, provides a good practical measure of prevention against the introduction of diseases along with imported aquatic species. Experience has shown that the 14 day quarantine period provides adequate time to allow fish that are incubating diseases or are carriers to become clinically affected (P. Beers, AQIS, Canberra, *pers. comm.*). However, because the system is based on detection of clinical signs of disease, it can be assumed that parasites, bacteria and viruses often enter the country. Humphrey *et al.* (1986) have shown that imported ornamental fishes originating from Southeast Asia carry latent bacterial infections exotic to Australia, while Anderson *et al.* (1993) found viral particles in dwarf gourami (*Colisa lalia*) imported from Singapore which died in Australian quarantine due to systemic amoebiasis. These authors recommend that inspection for clinical disease during quarantine should be extended to specific health accreditation or random sampling of all imported batches. The AQIS recognizes these shortcomings and in 1992 commissioned a major review of quarantine policy and the disease risks associated with importation of aquatic organisms and their products. The review has been completed and the report is to be released within the next few months (P. Beers, AQIS, Canberra, *pers. comm.*). The revised policy and quarantine system which will result from this effort should provide a useful model for quarantine development in the rest of the Asia-Pacific.

The status of quarantine and certification programs for Papua New Guinea has been summarized by ADB/NACA (1991) and by Kokolo (1995). Legislation concerning the import/export of aquatic species, including quarantine and certification procedures, is contained in the Animal Disease Control Act and the Fauna Protection Act (Kokolo, 1995).

Papua New Guinea is currently in the process of introducing a number of exotic species in the hope of improving fish production in the Sepik and Ramu River Basin. These include three cold-water species, copper mahseer (*Acrossocheilus hexagonolepis*), putitor mahseer (*Tor putitora*) and snow trout (*Schizothorax richardsonii*); and three warmwater species, tawas (*Barbodes gonionotus*), pirapatinga (*Piaractus brachypomus,* syn. *Colossoma bidens)* and sabalo (*Prochilodus lineatus)*. Introductions are made by importation of live eggs originating from a source where disease-free certification is available. Prior to shipment, the eggs are treated with iodine and upon arrival in PNG, they are kept in quarantine for three months. As PNG currently lacks expertise and facilities for quarantine and disease diagnostics, samples of live fish are sent to overseas laboratories (e.g., University of California at Davis, Universiti Pertanian Malaysia) for disease testing before they are released from quarantine.

Southeast Asia

The situation in Southeast Asia was recently up-dated by Arthur (1995). Three countries, Indonesia, Malaysia and Thailand, have recently made substantial advances.

Indonesia, the first Southeast Asian country to establish a quarantine and inspection service (see Shariff, 1987; Arthur and Shariff, 1991; Arthur, 1995), legislated provisions for a new fish quarantine inspection system in 1992. Staff for the quarantine service have been hired and have received preliminary training, a list of certifiable pathogens has been drawn up, and a *blue book* of approved diagnostic methods is in preparation. Of six main ports identified for quarantine stations, three (Jakarta, Medan and Pontianak) are still functioning as ports of entry. Importation is done under permit, and a certificate of health issued by the country of origin is required (ADB/NACA, 1991). Fish are supposed to be quarantined for a period of not less than one month, inspected, and if found infected with communicable disease, destroyed. However, due to lack of infrastructure, routine quarantine and inspection of imports have not yet been fully implemented and fish generally receive only visual inspection for signs of disease.

In 1984, Malaysia enacted legislation to control the import and export of fish and to establish quarantine and certification procedures (Ang *et al.*, 1989). The Department of Fisheries (DOF) is establishing four Fish Quarantine and Fish Health Centers under the Sixth Malaysia Plan (1991-1995) (Kechik and Abdulla, 1993). Centers near Subang International Airport, at Tampoi and at Glugor are in operation, while a fourth at Bukit Kayu Hitam will be functional in 1996 (M. Shariff, Universiti Pertanian Malaysia, Serdang, *pers. comm.*). Under the Seventh Malaysia Plan (1996-2000), additional centers will be constructed at the new Kuala Lumpur International Airport at Sepang, and at Tambak Johor and Labuan. The centers are to implement quarantine of imported aquatic organisms and also issue health certificates for exports at the request of the importing country. The DOF is in the final stages of drawing up rules for import/export and fish quarantine. It has also implemented an inspection program for the premises of individual importers and exporters.

Thailand first adopted legislation restricting the importation of aquatic organisms in 1953. This Royal Decree, subsequently amended in 1982 and made into law in 1983, prohibits the introduction of 176 fish species, 21 invertebrates, and 16 algae (Piyakarnchana, 1989). The Aquatic Animal Health Research Institute (AAHRI), established in 1993 and located in Bangkok, centralizes responsibilities for research on diseases of aquatic animals under one authority. Thailand does not require the inspection or quarantine of imported fishes (Tonguthai, 1987). The fishery's law is currently being revised by the Thai Department of Fisheries and will contain a provision requiring importers to obtain a statement issued by the authorized office of the country of origin attesting to the health status of imported living aquatic animals (K. Tonguthai, AAHRI, Bangkok, *pers. comm.*). Health certificates for shipments of exported fish are issued at the exporter's request, while hatcheries and holding facilities are occasionally inspected and certified at the request of importing countries, such as Australia.

Considerable fish health expertise also exists in Singapore and the Philippines. Singapore does not consider the importation of exotic species to be a major concern (Cheong *et al.*, 1983; Chou and Lam, 1989). However, Ng *et al.* (1993) have recently argued that appropriate legislation and quarantine measures are needed to protect its native fish fauna. Quarantine and sanitization procedures, established by the government, are routinely practiced on many farms, but are not compulsory. Ornamental fishes are not inspected upon import, but are examined for clinical signs of disease before export under a health accreditation scheme. As of April 1995, all fish health matters are the responsibility of the newly created Animal and Plant Health Inspection Division (APHID). Singapore plans to formalize a disease surveillance program which will include monitoring of diseases of economic importance (F. Chua, APHID, Singapore, *pers. comm.*). The Philippines, although having suffered the ravages of Epizootic Ulcerative Syndrome, does not regard establishment of a quarantine and certification program as a high priority. Indeed, as elsewhere, recent legislation designed to remove import restrictions which hamper international trade has worsened the situation (see Arthur, 1995).

Brunei, Kampuchea, Laos, Myanmar and Vietnam all have little expertise or infrastructure related to fish health (see ADB/NACA, 1991), and none of these countries is currently developing quarantine programs for aquatic organisms.

East Asia

No countries in East Asia have implemented programs for the routine quarantine and inspection of imported fishes. Although Japan has vast experience and expertise in fish health, there are no controls on the introduction of aquatic animals based on ecological considerations (Chiba *et al.*, 1989), and fish eggs and juveniles for aquaculture purposes can be freely imported (H. Wakabayashi, University of Tokyo, *pers. comm.*). No program for the routine quarantine and inspection of imported fishes exists (Kijima, 1990; Ogawa, 1990), however, there is a limited system for the quarantine of fry destined for aquaculture (H. Wakabayashi, *pers. comm.*). In addition, various recommendations and guidelines to avoid the introduction of fish pathogens from overseas and disease transfer between prefectures have been developed (ADB/NACA, 1991).

Fish health expertise is generally less developed in the other countries of East Asia. The *Import and export animal and plant quarantine regulations of the People's Republic of China* were implemented in 1982 (ADB/NACA, 1991), followed by other regulations dealing with the import and export of both terrestrial and aquatic animals. Although it has been reported that China follows a strict quarantine procedure when new species are introduced (see Tan and Tong, 1989), given the lack of expertise, infrastructure and detailed knowledge of the pathogens of imported aquatic animals, it is unlikely that this system is effective. The transfer of fish diseases from one region of the country to another is also considered a major problem (ADB/NACA, 1991).

Hong Kong considers strict quarantine to be unworkable, due the nature of the local aquaculture industry, which depends heavily on imported fry, the government's desire to promote free trade and the perception that a quarantine system would be unenforceable and not cost effective (ADB/NACA, 1991). Thus, Hong Kong has no laws related to the quarantine of imported fishes. Korea has some legislation aimed at preventing the import or export of diseased fishes (ADB/NACA, 1991), however there does not seem to be provision for routine quarantine and inspection. Taiwan lacks laws governing the importation of aquatic organisms (De Silva, 1989). To deal with the risk of disease importation, a health certificate is required from the importing country, as is a pre-inspection by a local veterinarian prior to importation, measures which can, at best, be only partially effective (ADB/NACA, 1991).

South Asia

The situation in South Asia is rather similar to that of East Asia. India, for example, has no legislation on the importation or introduction of aquatic organisms (De Silva, 1989) nor is there a proper quarantine procedure in place for imported fishes (Shetty *et al.*, 1989; ADB/NACA, 1991). The situation in Sri Lanka, Nepal, Pakistan and Bangladesh (Subasinghe and Balasuriya, 1987; De Silva, 1989; ADB/NACA, 1991) is much the same, some regulations dealing with the import/export of fish and/or fish products having been established, but no effective regulations or procedures for quarantine and inspection existing.

CONCLUSIONS

The effective control of the international spread of fish pathogens remains a distant and elusive goal for much of the Asia-Pacific, as it is elsewhere in many developed and developing countries. Effective national quarantine programs, coupled with a reliable, internationally agreed-upon program of certification of exported aquatic animals at source, would reduce the risk of disease introductions considerably. Given the high costs of disease, both to aquaculture and to artesanal fisheries, such a program would surely result in a net financial savings to the region. Efforts to prevent the international spread of fish diseases would be greatly aided if national governments were to severely restrict both the range of aquatic species imported and their numbers.

The progress made by such countries as Australia, Indonesia, Malaysia and Thailand to restrict the international spread of aquatic animal pathogens is laudable. Effective systems take considerable time, effort and resources, and protocols must be developed and modified based on regional experience. New trends towards more effective disease control include a more ecologically based management approach to introductions and transfers via the use of catchment zones (river drainage basins) for freshwater systems and biologically homogeneous coastal zones for marine systems, and the inclusion of crustaceans and molluscs in plans for disease prevention.

For many countries, lack of sufficient political will at the national level seems to be the major problem preventing an effective system to be established. Once national governments give the problem high priority, other constraints (e.g., lack of infrastructure, trained personnel, knowledge, etc.) will be more readily surmounted. In this regard, international organizations such as the OIE, NACA, SEAMEO-BIOTROP and FAO, and regional professional societies, notably the Asian Fisheries Society and its Fish Health Section, should continue to play a key role, highlighting the problem to policy makers and bringing together groups of experts who are capable of working towards solutions.

ACKNOWLEDGMENTS

I thank the following colleagues for providing information used in the preparation of this paper: Peter Beers, Dick Callinan, Frederic Chua, Kazuo Ogawa, Rohana Subasinghe, Mohamed Shariff and Kamonporn Tonguthai.

REFERENCES

ADB/NACA, 1991. Fish Health Management in Asia-Pacific. Report of a Regional Study and Workshop on Fish Disease and Fish Health Management. Asian Developm. Bank, Agricult. Dep. Rep. Ser. No. 1. Network of Aquaculture Centres in Asia-Pacific. Bangkok, 629 p.

Anderson, I.G., H.C. Prior, B.J. Rodwell and G.O. Harris, 1993. Iridovirus-like virions in imported dwarf gourami (*Colisa lalia*) with systemic amoebiasis. Austr. Veter. J. 70: 66-67.

Ang, K.J., R. Gopinath and T.E. Chua, 1989. The status of introduced fish species in Malaysia. p. 71-82. *In* S.S. De Silva. (ed.) Exotic Aquatic Organisms in Asia. Asian Fish. Soc. Spec. Publ. No. 3.

Anon., 1984. Guidelines for implementing the ICES code of practice concerning introductions and transfers of marine species. Intern. Counc. Explor. Sea, Coop. Res. Rep. No. 130, 20 p.

Arthur, J.R., 1987. (ed.) Fish Quarantine and Fish Diseases in South and Southeast Asia: 1986 Update. Asian Fish. Soc. Spec. Publ. No. 1, 86 p.

Arthur, J.R., 1995. Efforts to prevent the international spread of diseases of aquatic organisms, with emphasis on the Southeast Asian Region. *In* M. Shariff, J.R. Arthur and R.P. Subasinghe. (eds.) Diseases in Asian Aquaculture II. Proceedings of the 2nd Symposium on Diseases in Asian Aquaculture, 24-29 October, 1993, Phuket, Thailand. *Asian Fish. Soc.*, Fish Health Sect., Kuala Lumpur. (In press)

Arthur, J.R., and M. Shariff, 1991. Towards international fish disease control in Southeast Asia. Infofish Intern. 3/91: 45-48.

Bauer, O.N., and G.L. Hoffman, 1976. Helminth range extension by translocation of fish. p. 163-172. *In* L.A. Page. (ed.) Wildlife Diseases. Plenum Publ. Co., New York.

Carlton, J.T., 1993a. Overview of the issues concerning marine species introductions and transfers. p. 65-67. *In* M.R. Devoe. (ed.) Introductions & Transfers of Marine Species. Achieving a Balance Between Economic Development and Resource Protection. Proceedings of the Conference & Workshop, October 30 - November 2, 1991, Hilton Head Island, South Carolina. S.C. Seagrant Consortium.

Carlton, J.T., 1993b. Marine species introductions by ships' ballast water: an overview. p. 23-25. *In* M.R. Devoe. (ed.) Introductions & Transfers of Marine Species. Achieving a Balance Between Economic Development and Resource Protection. Proceedings of the Conference & Workshop, October 30 - November 2, 1991, Hilton Head Island, South Carolina. S.C. Seagrant Consortium.

Carlton, J.T., 1993c. An international perspective on species introductions: the ICES Protocol. p. 31-34. *In* M. R. Devoe. (ed.) Introductions & Transfers of Marine Species. Achieving a Balance Between Economic Development and Resource Protection. Proceedings of the Conference & Workshop, October 30 - November 2, 1991, Hilton Head Island, South Carolina. S.C. Seagrant Consortium.

Chamberlain, G.W., 1994. Taura syndrome and China collapse caused by new shrimp viruses. World Aquacult. 25: 22-25.

Chiba, K., Y. Taki, K. Sakai, and Y. Oozeki, 1989. Present status of aquatic organisms introduced into Japan. p. 63-70. *In* S.S. De Silva. (ed.) Exotic Aquatic Organisms in Asia. Proceedings of a Workshop on Introduction of Exotic Aquatic Organisms in Asia. Asian Fish. Soc. Spec. Publ. 3.

Cheong, L., R. Chou, R. Singh and Chao T.-M., 1983. Singapore, p. 47-63. *In* F.B. Davy and A. Chouinard. (eds.) Fish Quarantine and Fish Diseases in Southeast Asia. Report of a Workshop held in Jakarta, Indonesia, 7-10 December 1982. Intern. Developm. Res. Centre Publ. IDRC-210e.

Chou, L.M., and T.J. Lam, 1989. Introduction of exotic aquatic species in Singapore. p. 91-97. *In* S.S. De Silva. (ed.) Exotic Aquatic Organisms in Asia. Asian Fish. Soc. Spec. Publ. No. 3.

Davy, F.B., and A. Chouinard, 1983. (eds.) Fish Quarantine and Fish Diseases in Southeast Asia. Report of a Workshop held in Jakarta, Indonesia, 7-10 December 1982. Intern. Developm. Res. Centre Publ. IDRC-210e, 79 p.

Davy, B., and M. Graham, (eds.) 1979. Diseases of Fish Cultured for Food in Southeast Asia. Report of a Workshop held in Cisarua, Bogor, Indonesia, 28 November - 1 December 1978. Intern. Developm. Res. Centre Publ. IDRC-139e, 32 p.

De Silva, S.S., 1989. (ed.) Exotic Aquatic Organisms in Asia. Asian Fish. Soc. Spec. Publ. No. 3, 154 p.

Djajadiredja, R., T.H. Panjaitan, A. Rukyani, A. Sarono, D. Satyani and H. Supriyadi, 1983. Country Reports: Indonesia. p. 19-30. *In* F.B. Davy and A. Chouinard. (eds.) Fish Quarantine and Fish Diseases in Southeast Asia. Report of a Workshop held in Jakarta, Indonesia, 7-10 December 1982. Intern. Developm. Res. Centre Publ. IDRC-210e.

Hoffman, G.L., 1970. Intercontinental and transcontinental dissemination and transfaunation of fish parasites with emphasis on whirling disease (*Myxosoma cerebralis*). p. 69-81. *In* S.F. Snieszko. (ed.) A Symposium on Diseases of Fishes and Shellfishes. Am. Fish. Soc. Spec. Publ. No. 5.

Humphrey, J.D., 1986. Fish health, disease control and aquaculture. p. 107-114. *In* P. Owen and J. Bowden. (eds.) Freshwater Aquaculture in Australia. Rural Press Queensland Pty Ltd., Queensland.

Humphrey, J.D., C. Lancaster, N. Gudkovs and W. McDonald, 1986. Exotic bacterial pathogens *Edwardsiella tarda* and *Edwardsiella ictaluri* from imported ornamental fish *Betta splendens* and *Puntius conchonius*, respectively: isolation and quarantine significance. Austr.Veter. J. 63: 369-371.

Kechik, I. A., and S. Z. Abdulla, 1993. Fish quarantine in Malaysia. p. 61-69. *In* R.P. Subasinghe and M. Shariff. (eds.) Diseases in Aquaculture: the Current Issues. Malaysian Fish. Soc., Kuala Lumpur.

Kijima, T., 1990. Fish disease control plans in Japan. p. 971-972. *In* R. Hirano and I. Hanyu. (eds.) The Second Asian Fisheries Forum. Asian Fish. Soc., Manila.

de Kinkelin, P., C. Michel and P. Ghittino, 1985. Precis de Pathologie des Poissons. Institut National de la Recherche Agronomique / Office International des Épizooties, Paris, 348 p.

Kinne, O., 1984. Introduction to Volume IV, Part 1: Pisces. p. 1-16. *In* O. Kinne (ed.) Diseases of Marine Animals. Vol. IV, Part 1. Biologische Anstalt Helgoland, Hamburg, 541 p.

Kokolo, U., 1996. Quarantine practices used in Papua New Guinea for introductions and transfers of live fish. p. 140-142. *In* R.P. Subasinghe, J.R. Arthur and M. Shariff. (eds.). Health Management in Asian Aquaculture. Proceedings of the Regional Expert Consultation on Aquaculture Health Management in Asia and the Pacific. FAO Fish. Techn. Pap. 360.

Langdon, J.S., G.L. Enriquez and S. Sukimin, (eds.) 1992. Proceedings of the Symposium on Tropical Fish Health Management in quaculture. Biotrop Spec. Publ. No. 48, 172 p.

Lehane, L., 1992. Fish quarantine under scrutiny in 1993. Pet Indust. News, Australia 10: 15-16

Lightner, D.V., 1985. A review of the diseases of cultured penaeid shrimps and prawns with emphasis on recent discoveries and developments. p. 79-103. *In* Y. Taki, J.H. Primavera and J.A. Llobrera. (eds.). Proceedings of the First International Conference on the Culture of Penaeid Prawns/Shrimps. Iloilo City, Philippines, 1984. SEAFDEC Aquacult. Dep., Iloilo.

Lightner, D.V., 1990. Viroses section: introductory remarks. p. 3-6. *In* F.O. Perkins and T.C. Cheng. (eds.) Pathology in Marine Science. Academic Press, San Diego.

Lilley, J.H., M.J. Philips and K. Tonguthai, 1992. A review of epizootic ulcerative syndrome (EUS) in Asia. Publ. by Aquatic Animal Health Research Institute and Network of Aquaculture Centres in Asia-Pacific, Bangkok, 73 p.

Meyer, F.P., J.W. Warren and T.G. Carey, 1983. A Guide to Integrated Fish Health Management in the Great Lakes Basin. Great Lakes Fish. Comm. Spec. Publ. 83-2, 262 p.

Ng, P.K.L., L.M. Chou and T.J. Lam, 1993. The status and impact of introduced freshwater animals in Singapore. Biol.Conserv. 64: 19-24.

O'Connor, P.F., 1990. Introduced fish - the role of the Advisory Committee on Live Fish. p. 21-23. *In* D. A. Pollard. (ed.) Introduced and Translocated Fishes and the Ecological Effects. Australian Society for Fish Biology Workshop, Magnetic Island, 24-25 August 1989. Bur. Rural Resourc. Proc. No. 8.

Ogawa, K., 1990. Fish parasitology in Japan, its history and recent trends. p. 967-969. *In* R. Hirano and I. Hanyu. (eds.) The Second Asian Fisheries Forum. Asian Fish. Soc., Manila.

Piyakarnchana, T., 1989. Exotic aquatic species in Thailand. p. 119-124. *In* S.S. De Silva (ed.) Exotic Aquatic Organisms in Asia. Asian Fish. Soc. Spec. Publ. No. 3.

Porter, T.R., 1992. Protocols for the Introduction and Transfer of Salmonids. North Atlantic Salmon Conservation Organization, North American Commission, NAC (92) 24, 119 p.

Schmidt, M.E., and M.D. Love, 1985. The Australian quarantine system for live ornamental fish. Queensland Agricult. J., Jan.- Feb. 1985, p. 29-31.

Shariff, M., 1987. Current status of programs for fish health certification and quarantine systems. p. 48-52. *In* J.R. Arthur. (ed.) Fish Quarantine and Fish Diseases in South and Southeast Asia: 1986 Update. Asian Fish. Soc. Spec. Publ. No. 1.

Shariff, M., 1995. Fish health: an odyssey through the Asia-Pacific Region. Syarahan Inaugural, Univ. Pertanian Malaysia, Serdang, 25 p.

Shetty, H.P.C., M.C. Nandeesha and A.G. Jhingran, 1989. Impact of exotic aquatic species in Indian waters. p. 45-55. *In* S.S. De Silva. (ed.) Exotic Aquatic Organisms in Asia. Proceedings of a Workshop on Introduction of Exotic Aquatic Organisms in Asia. Asian Fish. Soc. Spec. Publ. 3.

Sindermann, C.J., 1986. Strategies for reducing risks from introductions of aquatic organisms: a marine perspective. Fisheries, 11 (2):10-15.

Sindermann, C.J., 1992. Disease risks associated with importation of nonindigenous marine animals. Mar. Fish. Rev. 54, No.3: 1-10.

Stewart, J.E., 1991. Introductions as factors in diseases of fish and aquatic invertebrates. Can. J. Fish. Aquat. Sci. 48: 110-117.

Subasinghe, R.P., and L.K.S.W. Balasuriya, 1987. Fish quarantine and certification in Sri Lanka. p. 73-76. *In* J. R. Arthur. (ed.) Fish Quarantine and Fish Diseases in South and Southeast Asia: 1986 Update. Asian Fish. Soc. Spec. Publ. 1.

Tan Y.-J and H.-Y. Tong, 1989. The status of exotic aquatic organisms in China. p. 35-43. *In* S.S. De Silva. (ed.) Exotic Aquatic Organisms in Asia. Proceedings of a Workshop on Introduction of Exotic Aquatic Organisms in Asia. Asian Fish.Soc .Spec. Publ. 3.

Tonguthai, K., 1985. A preliminary account of ulcerative fish diseases in the Indo-Pacific Region (a comprehensive study based on Thai experiences). National Inland Fisheries Institute, Bangkok, 39 p.

Tonguthai, K., 1987. Fish quarantine and certification in Thailand. p. 77-84. *In* J.R. Arthur. (ed.) Fish Quarantine and Fish Diseases in South and Southeast Asia: 1986 Update. Asian Fish. Soc. Spec. Publ. 1.

Turner, G.E. (ed.), 1988. Codes of practice and manual of procedures for consideration of introductions and transfers of marine and freshwater organisms. EIFAC Occas. Pap .No. 23, 44 p.

Williams, E.H. and C.J. Sindermann, 1992. Effects of disease interactions with exotic organisms on the health of the marine environment. p. 71-77. *In* M.R. Devoe. (ed.) Introductions & Transfers of Marine Species. Achieving a Balance Between Economic Development and Resource Protection. Proceedings of the Conference & Workshop, October 30 - November 2, 1991, Hilton Head Island, South Carolina. S.C. Seagrant Consortium.

Wootten, R., 1991. Legislation and the control of fish diseases. p. 37-38. *In* Fish Health Management in Asia-Pacific. Report of a Regional Study and Workshop on Fish Disease and Fish Health Management. Asian Developm. Bank, Agricult. Dep. Rep. Ser. No. 1. Network of Aquaculture Centres in Asia-Pacific, Bangkok.

Fish and Mollusc Health Research in the Asia-Pacific: Present Status and Future Directions

R.B. Callinan

NSW Fisheries Regional Veterinary Laboratory
Wollongbar 2477, Australia

Callinan, R.B. 1996. Fish and mollusc research in the Asia-Pacific: present status and future directions. *In* Health Management in Asian Aquaculture. Proceedings of the Regional Expert Consultation on Aquaculture Health Management in Asia and the Pacific. R.P. Subasinghe, J.R. Arthur & M. Shariff (eds.), p. 29-40. FAO Fisheries Technical Paper No. 360, Rome, FAO. 142 p.

Abstract

Fish health research in the region currently presents enormous challenges and opportunities. In many countries, resources for disease diagnosis and research are insufficient to meet the requirements of rapidly expanding aquaculture sectors. Relatively little is known about the full spectrum of infectious and non-infectious diseases affecting fish and molluscs in the region. It is essential that, for major industries, available research resources are focused on important disease-related production constraints with the aim of providing cost effective ways of reducing losses. Key issues and obstacles in the research process relating to identification of such production constraints, preparation of targeted funding applications, conduct of successful research projects and publication and extension of findings are discussed. To optimize research benefits, priority areas which should be addressed in most countries of the region include the need for more trained staff, particularly pathologists and epidemiologists, the need for improved access to scientific information and the need for improved diagnostic and research facilities.

INTRODUCTION

In many countries of the Asia-Pacific Region, the various aquaculture industries are now recognized as major forms of primary production having important economic, social and environmental impacts. The extent and intensity of regional aquaculture development, together with the usually unrestricted translocations of aquatic animals between and within countries, have led to frequent and serious outbreaks of disease. Many of these diseases are new or little understood. Moreover, in most countries of the region, the variety and pace of development of aquaculture industries have far outstripped the limited resources available for disease diagnosis and research. This situation presents both enormous challenges and opportunities to national and international agencies, to research-funding bodies and to individual researchers, diagnosticians, extension workers and aquaculturists. It is clearly essential that the limited resources available for disease diagnosis and research are used to maximum effect at regional and national levels. This paper utilizes the findings of a study of research needs for tropical aquaculture, conducted by the World Bank (1991), and a study of fish health management in the Asia-Pacific Region, conducted by the Asian Development Bank/Network of Aquaculture Centres in Asia-Pacific (1991). The paper also examines the processes whereby researchable issues are identified and, from the perspective of the senior researcher, examines aspects of funding application preparation and research project management in the current climate, with a view to identifying potential problem areas. Some practical measures to assist finfish and mollusc disease diagnosis and research are also suggested.

BACKGROUND

Since 1990, there have been three specific assessments of needs for fish disease research in tropical aquaculture:

I. 1991 World Bank Study

In considering the potential contributions of science to aquaculture, the World Bank (1991) recognized a major deficiency in knowledge of pathology (including knowledge of pathogens) of tropical fish and shellfish. The study identified the following research needs and opportunities in the field of tropical fish diseases:

Short-term:

- description of the clinical pathology of the tropical prawn, mollusc and finfish species;
- development of appropriate tissue cultures for tropical species as above;
- development of monoclonal antibody technology;
- development of nutritional pathology, particularly for fish meal substituted diets.

Long-term:

- development of disease-resistant strains of key species;
- development of comprehensive monitoring schemes of fish health and clinical monitoring system;
- development of specific pathogen-free cryo-preserved gene pools of gametes for transportation instead of adults, and to preserve the wild stock gene pools;
- development of environmentally safe drugs;
- development of bio-engineered vaccines against parasites and viruses;
- understanding, and exploitation, of the *specific* defense mechanisms of crustaceans and molluscs.

The study considered means whereby geographic disparities in knowledge could be overcome. It identified conditions necessary for conduct of efficiently targeted research, assessed institutional requirements and suggested a framework for inter-regional cooperation in tropical aquaculture research. The study concluded that networks involving core teams of scientists, networking organizations such as the Asian Fisheries Society, collaborative research programs and networks of research laboratories in developing and developed countries would accelerate development of tropical aquaculture research.

II. 1991 ADB/NACA Study

Requirements for diagnosis of, and research on, finfish and mollusc diseases were assessed within a major study of fish health management in the Asia-Pacific Region conducted by the Asian Development Bank and Network of Aquaculture Centres in Asia-Pacific (ADB/NACA 1991). The study acknowledged that the technical needs for disease diagnosis and research within the region are diverse, and that major distinctions exist between the requirements of technically advanced countries and the requirements of countries with less developed aquaculture production.

Regional approaches to strengthening diagnosis and research

In his paper in the ADB/NACA study document, Macintosh (1991) proposed the establishment of a regional fish health management resource unit within NACA. One of the purposes of this unit would be to provide supporting information to scientists in the region. In this regard, the unit would monitor regional fish disease status, operate a networked *early warning system*, source information from other regions, organize workshops/information exchange, facilitate donor agency coordination and operate an expert scientist registry for fish health management. Macintosh (1991) also proposed formation of regional expert committees for fields including diagnostics, research standardization, improved research methodology and provision of reference materials for standard diagnoses.

National approaches to strengthening diagnosis and research

On a national level, Macintosh (1991) suggested that most countries in the region needed more trained manpower, better research and diagnostic facilities and information support via *an early warning system*. In addition, countries with less developed aquaculture systems lacked equipment, capability to manage facilities, and manpower and expertise. They needed training and expert advice, as well as information on diagnosis and research. Countries with more developed aquaculture systems were seen as needing an *early warning system*, more research into the study of new diseases, more research on virology, nutritional diseases and environmental factors in disease, and more research on acceptable therapy and prevention methods.

In its summary assessment of national and regional needs for diagnosis and research, the ADB/NACA study identified the following as requiring strengthening:

Regional diagnostic needs

- Improved diagnostic facilities to provide a full range of services to aqua-farmers;
- greater on-farm technical support, involving better diagnostic capabilities in the field as well as in the laboratory;
- risk assessment of the use of chemicals and drugs for disease prevention or control;
- survey of regional aquatic disease status and standardization of diagnostic methods.

Regional research needs

- Upgrading the quality and quantity of research, particularly in the areas of virology, general microbiology, parasitology, clinical and general pathology, immunology, environmental chemistry, toxicology, fish nutrition, farm management and farming systems;
- greater attention in research to the on-farm situation and the environmental conditions which affect fish health;
- new research on the carrying capacity of environments which support aquaculture;
- research into drug and chemical uses in aquaculture;
- new research into the impact of diseases on mollusc culture.

Recommendations

The study recommended the following in relation to diagnosis and research:

- ***Establishment of a core group for coordination of national and regional programs***

Establishment of this core group at NACA was seen as the central action for development of fish health programs in the region. The group would be responsible for information collection and dissemination, coordination of cooperative research, establishing and maintaining regional centers of expertise and provision of assistance to national institutes.

- ***Upgraded national fish health management programs***

A national nodal center should be established in each country. The center would be the main diagnostic facility and would be linked to subnational centers for monitoring of disease status, diagnosis and training.

- *Establishment of a regional network*

The network would support national fish health management programs through dissemination of information on communicable diseases, cooperative research and provision of specialized assistance to national institutes as required.

- *Requirement for a regional center mechanism*

Regional centers of fish disease control, selected on the basis of commodity and discipline expertise, are required to support national centers in matters of diagnosis, technical reference and regional research.

- *Working groups to address specific issues*

Working groups are required to address specific issues, including standardization of fish disease diagnostic methods.

- *Requirement for international collaboration*

Collaboration is required with regional and international organizations in the development of management programs; the key role of NACA was recognized.

III. Second AFS/FHS Symposium on Diseases in Asian Aquaculture, Phuket, 1993 - Report of the session on *SIFR and the need for demand-led research,* held at the *Second Symposium on Diseases in Asian Aquaculture: Aquatic Animal Health and the Environment, Phuket, 25-29th October 1993*, organized by the Fish Health Section, Asian Fisheries Society (see Anon., 1993).

The report (see Anon., 1993) recognized that many factors contribute to the optimization of aquaculture production and that disease is one of the most serious constraints in this regard. The report had two stated objectives:

- To provide a research basis for short-term benefit to a rapidly growing and very diverse aquaculture industry;

- To provide a basis for stable long-term research which, although demand-led, allows basic and strategic needs to be met in addition, through sustained support over a long period.

General topics of importance in relation to constraint removal in Asian marine and freshwater finfish and shellfish culture were identified as :

- ***Assessment of the disease status of Asian aquaculture in both pathological and economic terms***

The report recommended that the impact of infectious diseases on aquaculture be measured by using standardized diagnostic methods in typical culture units to determine the identity and effects of diseases, and to estimate their economic costs in relation to mortality, production loss, feed conversion, etc.

- ***Basic studies on defense mechanisms and immunology of the many diverse vertebrate and invertebrate species farmed, and the pathogenesis of their disease processes***

The report recognized the narrow scientific base upon which the multibillion dollar aquaculture industries are based and expressed great concern at the lack of basic knowledge on host / pathogen / husbandry / environment interactions in almost all tropical aquaculture systems.

- ***Improvement in diagnostic techniques, particularly with regard to rapid diagnosis and certification***

The report recognized that equipment and skills necessary for disease diagnosis are not uniformly available, particularly in the poorer parts of the region. It recommended that conventional diagnostic capabilities be expanded. It also emphasized the need for rapid diagnostic methods such as fluorescent antibody and ELISA techniques, many of which remain to be developed for tropical aquaculture systems.

- ***Control of diseases by systems management***

The report recommended that management practices be developed to reduce environmental stress on the animals and reduce the prevalence and severity of disease. It suggested that development of new chemotherapeutic agents which are effective, do not lead to rapid antibiotic resistance and which leave minimal residues, is also required. In recognizing the concerns of many importing nations, the report recommended development of quality control procedures and methods for screening aquaculture products for pathogens and chemical residues.

- ***Environmental impact***

The report recognized the key role played by environmental factors in infectious disease causation and recommended studies to evaluate pathogen/host/environment relationships. It suggested that, by improving aquaculture management practices, stressful conditions can be reduced, thus improving survival, growth and performance.

- *Training*

The report recognized that research results must be effectively communicated to end users. It recommended that mechanisms for technology transfer be improved. The peer-review journal system was seen as central to such technology transfer. Mechanisms for technology transfer from scientists in lead centers, both within and outside the region, to scientists working in less developed centers were identified as very important. In addition, mechanisms for transfer of knowledge and skills to extension trainers and to farmers were seen as critical.

CURRENT ACTIVITIES IN REGIONAL FISH HEALTH RESEARCH AND DIAGNOSIS

AAHRI and the South East Asian Aquatic Disease Control Project

In 1991, the Aquatic Animal Health Research Institute (AAHRI) was established in Bangkok as the base for the South East Asian Aquatic Disease Control Project, jointly funded by British Overseas Development Administration and the Thai government. The Project aims to develop AAHRI as a regional lead center through programs designed to identify and meet regional needs in diagnosis and research, as well as by disseminating information. Importantly, the Project provides opportunities for researchers to train via participation in research programs at AAHRI. Currently, Malaysia, Indonesia, Nepal, Bangladesh, Cambodia, Vietnam, Laos, Philippines and Thailand are involved. The Project also plans to conduct independently funded research projects based at AAHRI. These may include collaborative studies involving institutions within or beyond the region. Annual Network Workshops, at which research, current disease problems, training requirements and other development constraints will be discussed, are also planned.

Collaborative Research Projects

Donor agencies such as Australian Center for International Agricultural Research (ACIAR) and Japan International Research Center for Agricultural Sciences (JIRCAS) provide funding for collaborative research projects in fields including fish health. Currently, ACIAR is funding a project on epizootic ulcerative syndrome involving researchers from Australia, India, Indonesia and the Philippines. Such programs make important contributions to meeting many of the needs identified for fish health research in the region.

FURTHER DEVELOPMENT OF REGIONAL RESEARCH AND DIAGNOSTIC ACTIVITIES

While much is known about individual disease-causing organisms, relatively little is known about the full spectrum of infectious and non-infectious diseases seen in fish, and we are still at the relatively basic level of properly defining the full range of syndromes (ADB/NACA 1991). Under these circumstances, research and diagnosis on finfish and mollusc health are obviously closely linked and will be discussed together in this paper.

Given the limited knowledge of fish diseases in most countries of the region, it is essential that diagnostic and research activities in the short to medium term are "demand led" and directed at identifying those diseases causing the greatest production losses and at developing sustainable control and prevention measures for them.

Steps in implementing a research program on key production constraints

The procedure outlined below represents an ideal which, even in developed countries, is often not completely achievable. Important components are currently lacking in many developing countries of the region. The procedure is presented here so that areas of need at national or regional levels can be specifically identified. The procedure could also assist senior researchers in refinement of methods currently used to identify appropriate research topics, to obtain funding and to conduct fish health research programs.

Identification and ranking of diseases/syndromes causing key production constraints in finfish and mollusc culture

In recent decades, massive production losses in aquaculture in the Asia-Pacific Region have been caused by diseases such as epizootic ulcerative syndrome (EUS) and pearl clam mortalities (ADB/NACA 1991). Losses caused by these diseases were recognized as so severe that research into control and prevention methods was immediately justified without detailed comparisons with losses caused by other diseases being necessary. However, it is highly likely that other less spectacular outbreaks of infectious and non-infectious diseases also cause significant, but usually unrecorded, production losses in regional aquaculture. It is also likely that, as national quarantine regulations are improved and enforced, research attention will turn to control and prevention of important endemic diseases. These must first be identified, and the necessary information can be gathered from at least three levels:

- Fish farmers, in consultation with extension workers, are essential sources of information. They can provide information on occurrence of easily recognized specific diseases (*e.g.*, white spot, lymphocystis), on easily recognized syndromes (*e.g.*, cutaneous ulcers, skeletal deformities) and on poorly defined syndromes (*e.g.*, suboptimal growth rates, non-specific mortalities, new diseases). Laboratory diagnosis is required for accurate identification and characterization of diseases in the latter two categories.

- Diagnostic laboratories can be a reliable source of information on disease occurrence and distribution. However, to accurately identify any disease, laboratories must be staffed by adequately trained scientists. As a general signpost to the problems affecting a population at risk, the discipline of pathology has no parallel. Performing necropsies is the most fruitful approach in solving fish health problems (ADB/NACA, 1991). Therefore, as an essential minimum, diagnostic laboratories servicing aquaculture must be staffed by at least one diagnostician with expertise in pathology, including histopathology; appropriate technical support in histopathology must also be available. Scientific and technical staff with expertise in support disciplines including bacteriology, parasitology, mycology and water quality are also necessary for accurate identification of diseases at a diagnostic laboratory.
- Over time, a diagnostic laboratory accumulates information which can be used to identify the relative importance, within the area serviced, of any disease as a production constraint. As part of this identification process, allowance must be made for factors which influence submission patterns, *e.g.*, the increased difficulty of submitting suitable material from more distant parts of the area serviced.
- Specialist laboratories may be available to provide support to diagnostic laboratories in disciplines such as virology, immunology, biochemistry, analytical chemistry and electron microscopy. They therefore also provide important information which can be used to identify those diseases causing important production constraints.

For each disease, information indicating its relative importance as a demand-led production constraint can be accumulated over time from the above sources and through close consultation with industry bodies and farmer representatives. Senior researchers, economists, epidemiologists and industry representatives can then rank the diseases using economic, environmental and sociological criteria. Often *educated guesses* must be used to estimate economic costs. It is important that such estimates are arrived at as objectively as possible.

Key researchable aspects of important diseases can then be identified and potential benefits arising from such research assessed. The cost of a key production constraint, the cost of the proposed research program and the cost of the expected control program must be evaluated against the likely economic benefits of the control program. Objectivity, rather than researchers' preferences and interests, should likewise dominate this process.

Preparation of a targeted funding proposal

Once a researchable topic has been identified, in many cases external funds will be required before a project can proceed. The preparation of a funding application is a critical and demanding process and, in many developing countries, necessary information is often not readily available. Such information includes lists of potential donor bodies, so that the application can be effectively targeted. To minimize the risk of duplication and to maximize the benefits of the program, the funding application must be based on a comprehensive and detailed literature review. Access to scientific literature may be limited or difficult to obtain, so that necessary technical information may be lacking. Competence in English and Japanese, the major languages of fish health literature, or access to competent translators, is often also limited.

Design and conduct of the research project

Access to current and past scientific literature is essential. It helps ensure that the separate experiments within a research project are designed as logical extensions of existing knowledge and that they use appropriate technical methods. Competence in English and Japanese may be a limiting factor in this regard for some workers in the region.

Much work can be wasted if experimental design is flawed. Before formal experimental protocols are prepared, it is essential that detailed consultations are held with biometricians or others with expertise in experimental design to ensure the experimental outcomes will be valid and useful.

Adequate scientific and administrative infrastructure is obviously essential if the research is to succeed. For example, in a project with an essential histopathology component and involving an experienced pathologist, major problems may emerge if technical facilities are inadequate, key consumables are difficult to obtain or the laboratory histopathology technician is inexperienced.

During the conduct of the experiments themselves, unexpected results and unforeseen problems will almost certainly occur. Access to literature is again important at this stage, as is access to experienced senior researchers who can advise and motivate. These may not be available in some developing countries of the region.

Publication of results in peer-reviewed international journals

Research results must be communicated, as appropriate, to producers, extension workers, other scientists and to government agencies. This is often done effectively within countries of the region, but there is a clear need to improve international dissemination of results. For a variety of reasons, some of which are mentioned below, much of the research on fish health in the region is never published in peer-reviewed international journals and, as a result, some of the research benefits are lost. Successful preparation of a suitable manuscript is a demanding process with many potential obstacles. An ability to write clearly and concisely in English is essential. Preparation also requires access to scientific literature, access to computerized word processing, data analysis and graphics applications. Often, specialized technical equipment, such as access to a well maintained, functional photomicroscope, is also required. Assistance in these areas is required in many countries of the region.

DISCUSSION

The recommendations of the World Bank (1991) and ADB/NACA (1991) studies, as summarized above, remain substantially valid and will continue to provide a basis for development of diagnostic and research capabilities in fish health for the region. Generally, this development is proceeding, albeit slowly, and the starting points and rates of progress vary widely between countries. The following modifications and additions to the recommendations are suggested as practical means to assist development of diagnostic and research capabilities in finfish and mollusc health during this transition period:

Additional Recommendations

- Diagnostic and research effort should be demand-led and focused on diseases causing major production constraints.
- Because pathology plays a pivotal role in all fish disease diagnosis and in much disease-related research, appropriate training programs are urgently required to correct an acute shortage of skilled pathologists in many countries of the region.
- Information support to diagnosticians and researchers is urgently required and should include ready access to scientific literature and direct access to case reference material (including kodachromes and histological sections illustrating typical cases).
- An information package on all funding agencies which may support work in fish disease diagnosis and research in the region should be made available to appropriate senior researchers in each country. The package, which should be updated as necessary, should list all relevant agencies, and include information on funding policy, formats and deadlines for applications.
- It is essential that expert advice on experimental design is available to researchers so as to maximize the benefits of their work.
- Assistance with preparation of papers for publication in English is urgently required by many diagnosticians and researchers and would help maximize the benefits of their work. Excellent manuals are available on this topic and should be made available as necessary. Where appropriate, direct assistance may be warranted.

ACKNOWLEDGMENTS

I gratefully acknowledge the information and assistance provided during the preparation of this paper by J.R. Arthur, G. Barua, T. Hideyuki, S. Lumanlan-Mayo, I.H. Macrae, T.

Miyazaki, C.V. Mohan, M.J. Phillips, C.R. Lavilla-Pitogo, B. O'Neil, M. Shariff and R.P. Subasinghe.

REFERENCES

ADB/NACA, 1991. Fish Health Management in Asia-Pacific. Report on a Regional Study and Workshop on Fish Disease and Fish Health Management. ADB Agricult. Dep. Rep. Ser. No. 1. Network of Aquaculture Centres in Asia Pacific. Bangkok, Thailand.

Anon., 1993. Priorities for disease research: towards minimizing losses from disease in Asian Aquaculture. Asian Fish. Soc., Fish Health Sect. Newsletter 4(2): 4-5, 10.

Macintosh, D.J., 1991. Research and Training Requirements and Opportunities. *In* Fish Health Management in Asia-Pacific. Report on a Regional Study and Workshop on Fish Disease and Fish Health Management. ADB Agricult. Dep. Rep. Ser. No. 1. Network of Aquaculture Centres in Asia Pacific. Bangkok, Thailand. p. 549-567.

The World Bank/UNDP/CEC/FAO, 1991. Tropical Aquaculture Development: Research Needs. World Bank Tech. Pap. No. 151. Fish. Ser. The World Bank. Washington, D.C.

Shrimp Health Research in the Asia-Pacific: Present Status and Future Directives

Celia R. Lavilla-Pitogo

Fish Health Section, Aquaculture Department,
Southeast Asian Fisheries Development Centre
Tigbauan, Iloilo, Philippines

Lavilla-Pitogo, C.R. 1996. Shrimp health research in the Asia-Pacific: present status and future directives. *In* Health Management in Asian Aquaculture. Proceedings of the Regional Expert Consultation on Aquaculture Health Management in Asia and the Pacific. R.P. Subasinghe, J.R. Arthur & M. Shariff (eds.), p. 41-50. FAO Fisheries Technical Paper No. 360. Rome, FAO. 142 p.

Abstract

Shrimp harvests from intensive aquaculture have recently declined in areas which have been productive for many years. Because the most convenient, although not necessarily factual, explanation for these crop failures has been the occurrence of infectious diseases, there is a need to consider shrimp health from a holistic point of view. The classical method used for the study of shrimp disease dealt mainly with identification of the causative organism and the search for methods of prevention and control through chemotherapy. The adverse affects resulting from the use of chemicals in aquaculture have led to a clamor for alternative approaches to disease management. For effective shrimp health maintenance and surveillance, the following components need consideration: development of rapid and sensitive methods for pathogen detection; establishment of shrimp tissue cultures for virology; immunological studies, toxicological studies and drug efficacy evaluation. The epidemiological approach to disease management should augment the classic approach to shrimp pathology, and this calls for multidisciplinary cooperation.

INTRODUCTION

In 1991 and 1992, over 700 000 metric tonnes (mt) of cultured shrimp were produced worldwide, with 80% of production coming from Asia. Production was expected to increase in succeeding years with the opening of new farms and adoption of more intensive culture methods. This expectation was not met, however. Global shrimp production in 1993 was only 639 000 mt, about 12% lower than the 1992 production. Crop failure in China in the summer of 1993 greatly affected the world's farmed shrimp production. The increase in production seen in 1994 was mainly due to record crops in Thailand, India and Vietnam (Rosenberry, 1994). This production, however, still did not surpass the harvests obtained in 1991 and 1992. Table 1 shows the production records of major shrimp-producing countries in Asia from 1991-1994.

Black tiger shrimp, *Penaeus monodon*, is the main species cultured in Asia except in China and Japan, where the main crops are *P. chinensis* and *P. japonicus*, respectively. Other penaeids grown in farms are *P. semisulcatus*, *P. penicillatus*, *P. merguiensis* and *P. indicus*. The phenomenal growth of the shrimp farming industry in less than two decades has created various problems. Disease epizootics of economic significance are a major constraint to the industry, not only because they affect the quantity of harvest, but because disease also affects the quality and regularity of production. At present, the main goal of the shrimp industry is to meet the growing demand in a sustainable manner without damaging the environment. The role that shrimp disease research must play to attain such a goal is now being seriously considered by both the private and government sectors, and by national and international organizations. It is clear that collaborative effort involving all sectors is needed to attain sustainability. However, a sustainable industry can only be achieved if considerable investment in research is provided by many sectors.

MAJOR HEALTH PROBLEMS IN SHRIMP CULTURE FACILITIES IN THE ASIA-PACIFIC

Epizootics of both infectious and non-infectious etiology have continuously plagued the various sectors of the industry. Although it is generally recognized that intensive culture systems often encounter serious disease problems, more recent experiences have shown that low-density culture systems can also be severely affected.

A recent review on shrimp diseases has been made by Lightner (1993). In the context of Asian shrimp culture, excellent papers on diseases were presented at the *Workshop on Diseases of Cultured Penaeid Shrimp in Asia and the United States* (Fulks and Main, 1992). Diseases of both infectious and non-infectious etiology have been described, but their effects on shrimp and impacts on culture activities remain poorly understood. Diseases caused by viruses, bacteria, fungi and protistans are considered very significant to shrimp culture. The following list is largely based on Lightner (1993) and Lightner *et al.* (1994).

Diseases Due to Infectious Organisms

- ***Viruses***

At least 15 viruses are known to infect cultured and wild marine penaeid shrimp. Reported types include parvoviruses, baculoviruses, reoviruses, togaviruses and rhabdoviruses. Three systemic baculoviruses have recently been described for penaeid shrimp: yellowhead virus (YBV) (Boonyaratpalin *et al.*, 1993), hemolymph baculovirus (Owens, 1993) and systemic ectodermal and mesodermal baculovirus (SEMBV) (Wongteerasupaya *et al.*, 1995). YBV and SEMBV infections have caused drastic mortalities resulting in severe economic losses in shrimp culture facilities in Thailand, Indonesia and India. Also described is a rod-shaped nuclear virus of *Penaeus japonicus* (RV-PJ) (Inouye *et al.*, 1994) which has been implicated in mass mortalities of cultured *P. japonicus* in Japan in 1993.

- ***Bacteria***

Due to the economic losses following epizootics, bacteria are considered the most economically significant disease agents of shrimp. A review of the bacterial diseases affecting different aquaculture commodities in Asia is presented by Lavilla-Pitogo (*in press*). Infections due to *Vibrio* spp. affect shrimp in hatchery and growout facilities, causing chronic or mass mortalities and shell deformities which affect marketability. Although bacteria are easily isolated from diseased shrimp, the mechanism for pathogenicity remains unclear. The course of action taken by most farmers to arrest bacterial infections is to use antibacterials. These products have been used indiscriminately, creating resistance problems. Issues, problems and constraints regarding the use of chemicals in aquaculture facilities were raised during a workshop at the *Symposium on Diseases in Asian Aquaculture* held in Bali, Indonesia in 1990 (Shariff *et al.*, 1992).

- ***Fungal, protistan and other parasitic diseases***

Fungi like *Lagenidium*, *Haliphthoros* and *Sirolpidium* caused epizootics in shrimp hatcheries (Baticados *et al.*, 1990), resulting in heavy mortalities in larval stocks within two days. Infections due to the parasitic microsporidian *Agmasoma penaei* have been reported to cause high mortality in cultured shrimp (Flegel *et al.*, 1992). Protistans, such as gregarines and ciliates, are also associated with shrimp, but their general impact is less because of their relatively slow development.

Non-Infectious Diseases and Syndromes

Several nutritional diseases and syndromes have been reported in shrimp, such as chronic soft-shell syndrome (Baticados *et al.*, 1986) and red disease (Lightner and Redman, 1985). Various nutritional, toxic and environmental diseases are discussed in Sindermann and Lightner (1988), Baticados *et al.* (1990), Flegel *et al.* (1992) and Limsuwan (1993).

Reports of mortalities due to toxins and pesticides are emerging. Flegel *et al.* (1992) describe the toxic effects of methyl-parathion and cypermethrin pesticides used in rice culture on postlarvae of *Penaeus monodon*. The emphasis placed on studying these pesticides is due to the fact that they can find their way into shrimp farms through run-off.

Environmental Problems

Problems resulting from the effects of shrimp culture on the natural environment have been raised (Primavera, 1993). Moreover, concern is also directed toward the effects of shrimp culture practices on the pond environment and towards the conflict among users of the same water source for various purposes. More detailed discussions are given by Phillips (this volume).

THE IMPORTANCE OF RESEARCH

In the past, high production returns from shrimp farming were recorded in areas where new farms with good water source were opened. Opening of new areas is now limited, however, and future production will largely depend on already utilized resources. Successful production ventures will now largely depend on effective management utilizing technologies developed through meticulous research.

As with fish pathology, the classic approach to shrimp pathology is descriptive in nature. Most published studies on shrimp disease deal with the isolation and morphological or biochemical characterization of the pathogen. Recently, these activities have been carried out through molecular methods using gene probes. For effective disease control and surveillance programs, the following components should be considered:

Diagnostics

The use of new gene probe technologies that rely on demonstrating specific nucleic acid sequences offers an opportunity to detect viruses and other pathogens at much earlier stages of infection and to study how particular pathogens are transmitted to and within the shrimp host. Shrimp health status certification can also be performed with greater confidence in the outcome.

Epidemiology

The study of the distribution and determinants of health and disease in the natural setting is analogous to the pathogenesis of disease in individuals (Meek and Martin, 1991). It is clear that the classic approach to fish pathology should be augmented by epidemiology. Epidemiological surveys, which are essential for managing the spread of shrimp disease agents, need urgent attention. The availability of standardized diagnostic methods with predictable results will make surveillance and diagnostic work easier. The epidemiological approach calls for cooperation between farmers and pathologists in data-gathering to assess situations pre-empting a disease outbreak. These data will be the basis for rational decisions for the prevention and/or control of disease.

Tissue Culture Establishment

Although this work is expensive and difficult, cell lines are important because they allow *in vitro* study of host-pathogen interactions.

Immunologic Studies

For effective diagnosis and control of infectious shrimp diseases, sufficient knowledge of immunology is needed. The latest information and recommended research strategies for marine molluscan and crustacean immunology are discussed by Bachere *et al.* (1995). They emphasize the importance of immunology as a tool for studying host-pathogen interactions at the cellular and molecular levels, and as a key to selection of pathogen-resistant animals.

Toxicology and Drug Efficacy Evaluation

The need for approved chemotherapeutants in shrimp culture is obvious. However, the approval process required is complicated, time-consuming, and expensive. Bell (1992) thoroughly discusses the issues related to drugs for shrimp diseases. Equal to the need for approval is effective enforcement of laws regarding the use of drugs in shrimp culture.

Collaborative efforts with other disciplines will also lead to better shrimp health management. Husbandry techniques to produce fry and broodstock of good quality, as well as better nutrition and feeding strategies, will greatly support hatchery and farming activities. On the farm, environmentally friendly water and soil disinfection strategies are needed to get rid of pathogens in the water source from over-subscribed areas. Effluent clean-up schemes need to be devised to reduce nutrient discharge and prevent eutrophication in receiving waters. Mialhe *et al.* (1995) also emphasize that beyond the strategy of prophylaxis based on diagnostics for preventing and managing epidemics, another strategy must be developed for selecting pathogen-resistant strains.

The Importance of Collaboration in Shrimp Health Research

As a field of research, shrimp pathology has limited manpower compared to disciplines like breeding, seed production and husbandry. Solutions to production constraints due to infectious and non-infectious disease agents are not easily met, thus creating a feeling of uncertainty among investors.

It is clear that to solve enormous problems with limited manpower capability, there is a need for cooperation and collaboration among institutions. The concept of an "international team" to work on problems related to infectious diseases in marine invertebrate aquaculture was put forward by Mialhe *et al.* (1995) to avoid duplication of work.

There is also a need for increased interdisciplinary research. Pathologists can work with nutritionists in determining the nutritive values of new feed ingredients or the effects of anti-nutritional factors in these substances.

The role of scientific societies in carrying out plans for collaborative research is also discussed by M. Shariff (this volume). In Asia, the Fish Health Section of the Asian Fisheries Society serves as an effective mechanism for the exchange of ideas in research. However, private sector participation in this exchange is minimal. Research institutions, universities, government agencies and some private companies have contributed significantly to shrimp health research in Asia. Further strengthening of collaborative mechanisms to focus on shrimp diseases is still necessary to find fast track solutions to problems that beset the industry.

Effective collaborative projects have been initiated at the Aquatic Animal Health Research Institute (AAHRI) in Bangkok, Thailand with funding assistance from the Overseas Development Administration of the United Kingdom. The International Development Research Centre of Canada (IDRC) also funded research projects on shrimp disease at the Southeast Asian Fisheries Development Center (SEAFDEC) and the Philippine Bureau of Fisheries and Aquatic Resources (BFAR).

The role of universities in shrimp research is effectively carried out through the training of graduate students. This mechanism has boosted the manpower capability in shrimp pathology in Asia. Many universities and institutions in Japan, Taiwan, Malaysia, Australia and Thailand have contributed in training manpower in shrimp health research.

RECOMMENDATIONS FOR THE FUTURE

It is important for research to come up with information to secure the stability, sustainability and profitability of the shrimp industry. This could be done through:

- developing rapid and sensitive methods to detect pathogens within the animal and in its environment;
- increasing the availability and accessibility of disease diagnosis, prevention, and treatment services;
- increasing the availability of quality fry for stocking (fry which are disease resistant, fast growing, tolerant to sub-optimum environmental conditions etc.);
- developing a reliable method to assess fry quality;

- increasing the availability of environmentally sound and cost-effective culture methods (determination of pond carrying capacity, fine tuning of feeding strategies, waste recycling methods, polyculture etc.);
- undertaking studies on pond dynamics and its influence on shrimp health; and
- developing or verifying technologies for the use of probiotics, bioaugmentation and bioremediation products for shrimp aquaculture.

Sufficient resources to develop and transfer technologies to support an industry faced with huge problems should be made available. Government funds, as well as contributions from international funding agencies, have initially funded research. Closer interaction with industry representatives should be made in order to come up with schemes to help finance expensive research.

REFERENCES

Bachere E., E. Mialhe, D. Noel, V. Boulo, A. Morvan and J. Rodriguez., 1995. Knowledge and research prospects in marine mollusc and crustacean immunology. Aquaculture, 132: 17-32.

Baticados M.C.L., R.M. Coloso and R.C. Duremdez, 1986. Studies on the chronic soft-shell syndrome in the tiger prawn, *Penaeus monodon* Fabricius, from brackishwater ponds. Aquaculture, 56: 271-285.

Baticados M.C.L., E.R. Cruz-Lacierda, M.C. de la Cruz, R.C. Duremdez-Fernandez, R.Q. Gacutan, C.R.Lavilla-Pitogo and G.D. Lio-Po, 1990. Diseases of Penaeid Shrimps in the Philippines. Aquaculture Department, SEAFDEC, Tigbauan, Iloilo, Philippines, 46 p.

Bell T.A., 1992. Drugs and chemotherapeutants for shrimp diseases: their present status in the United States, with an overview of research and approval processes, p. 311-324. *In* W. Fulks and K. Main (eds.) Diseases of Cultured Penaeid Shrimp in Asia and United States. The Oceanic Institute, Honolulu, HI

Boonyaratpalin S., K. Supamataya, J. Kasornchandra, S. Direkbusarakom, U. Ekpanithanpong and C. Chantanachookin, 1993. Non-occluded baculo-like virus the causative agent of yellow-head disease in black tiger shrimp *Penaeus monodon*. Gyobo Kenkyu (Fish Pathol.) 28: 103-109.

Flegel T.W., D.F. Fegan, S. Kongsom, S. Vuthikomudomkit, S. Sriurairatana, S. Boonyaratpalin, C. Chantanachookhin, J.E. Vickers and O. D. MacDonald, 1992. Occurrence, diagnosis, and treatment of shrimp diseases in Thailand, p. 57-112. *In* W. Fulks and K. Main (eds.). Diseases of Cultured Penaeid Shrimp in Asia and United States. The Oceanic Institute, Honolulu, HI

Fulks, W., and K. Main (eds.), 1992. Diseases of Cultured Penaeid Shrimp in Asia and the United States. The Oceanic Institute, Honolulu, HI, 392 p.

Inouye, K., S. Miwa, N. Oseko, H. Nakano, T. Kimura, K. Momoyama and M. Hiraoka, 1994. Mass mortalities of cultured Kuruma shrimp *Penaeus japonicus* in Japan in 1993: electron microscopic evidence of the causative virus. Fish Pathol. 29: 149-158.

Lavilla-Pitogo, C.R., (in press). Bacterial diseases of penaeid shrimp: an Asian view. *In* M. Shariff, J.R. Arthur and R.P. Subasinghe (eds.) Diseases in Asian Aquaculture II. Fish Health Section, Asian Fish. Soc., Manila, Philippines.

Lightner, D.V., 1993. Diseases of penaeid shrimp, p. 393-486. *In* J.P. McVey (ed.) CRC Handbook of Mariculture. 2nd Edn., Vol. 1. Crustacean Aquaculture, CRC Press, Boca Raton, FL

Lightner, D.V., B.T. Poulos, L. Bruce, R.M. Redman, L. Nunan, C. Pantoja, J. Mari and J.R. Bonami, 1994. Development and application of genomic probes for use as diagnostic and research reagents for the penaeid shrimp parvoviruses IHHNV and HPV and the baculoviruses MBV and BP. USMSPF 10th Ann. Rev., Gulf Coast Res. Lab. Spec. Publ. No. 1, p. 59-85.

Lightner, D.V., and R.M. Redman. 1985. Necrosis of the hepatopancreas in *Penaeus monodon* and *P. stylirostris* with red disease. *J. Fish Dis.* 8: 181-188.

Limsuwan, C., 1993. Diseases of black tiger shrimp, *Penaeus monodon* Fabricius, in Thailand, p. 1-22. *In* D. Akiyama (ed.) Am. Soybean Soc.Tech. Bull., Singapore.

Meek, A.H., and S.W. Martin, 1991. Epidemiology and infectious disease, 141-180 p. *In* J.B. Woolcock (ed.) Microbiology of Animals and Animal Products. Elsevier Science Publ., Amsterdam.

Mialhe, E., E. Bachere, V. Boulo and J.P.Cadoret, 1995. Strategy for research and international cooperation in marine invertebrate pathology, immunology and genetics. Aquaculture, 132: 33-41.

Owens L., 1993. Description of the first haemocytic rod-shaped virus from a penaeid prawn. Dis. Aquat. Org. 16: 217-221.

Primavera J.H., 1993. A critical review of shrimp pond culture in the Philippines. Rev. Fish. Sci. 1: 151-201.

Rosenberry, B. (ed.), 1994. World Shrimp Farming 1994. Shrimp News International, San Diego, CA, 68 p.

Shariff, M., R.P. Subasinghe and J.R. Arthur (eds.), 1992. Diseases in Asian Aquaculture I. Fish Health Section, Asian Fish,. Soc., Manila, Philippines, 587 p.

Sindermann, C.J., and D.V. Lightner (eds.), 1988. Disease Diagnosis and Control in North American Marine Aquaculture. Developments in Aquaculture and Fisheries Science, Vol. 17. Elsevier Science Publ., Amsterdam, 431 p.

Wongteerasupaya C., J.E. Vickers, S. Sriurairatana, G.L. Nash, A. Akarajamorn, V. Boonsaeng, S. Panyim, A. Tassanakajon, B. Withyachumnarnkul and T.W. Flegel. 1995. A non-occluded, systemic baculovirus that occurs in cells of ectodermal and mesodermal origin and causes high mortality in the black tiger prawn *Penaeus monodon*. Dis. Aquat. Org. 21: 69-77.

Table 1. Production record of major shrimp farming countries in Asia (1991-1994).[1,2]

Countries	Head-on Production (metric tonnes x 1000)			
	1994	1993	1992	1991
Thailand	225	209	163	153
Indonesia	100	100	130	140
India	70	55	45	35
China	35	30	140	145
Vietnam	50	40	35	30
Bangladesh	35	30	25	25
Philippines	30	25	25	30
Taiwan	25	20	25	30
Total	**570**	**509**	**588**	**588**

[1]1994 figures are from Rosenberry (1994); 1991-1993 figures are from C.P. Shrimp News (March 1994).

[2]Other countries which also produce pond-grown shrimp include Japan, Korea, Sri Lanka, Australia and Malaysia.

Role of Non-Governmental Organizations (NGOs) in Fish Health Management in the Asia-Pacific

Mohamed Shariff

Faculty of Veterinary Medicine, Universiti Pertanian Malaysia,
43400 Serdang, Selangor, Malaysia

Shariff, M. 1996. Role of non-governmental organizations (NGOs) in fish health management in the Asia-Pacific. *In* Health Management in Asian Aquaculture. Proceedings of the Regional Expert Consultation on Aquaculture Health Management in Asia and the Pacific. R.P. Subasinghe, J.R. Arthur & M. Shariff (eds.), p. 51-59. FAO Fisheries Technical Paper No. 360, Rome, FAO. 142 p.

Abstract

NGOs are self-governing, independently financed organizations and societies that assign themselves specific roles, issues or disciplines paralleling governmental or intergovernmental organizations that have similar mandates. In Asia, regional NGOs have been quite active in promoting sustainable fisheries and aquaculture. In particular, the Asian Fisheries Society (AFS) has been involved in this activity for more than a decade, and its Fish Health Section (FHS/AFS) is an NGO with a clear mandate to promote aquaculture health management in the region, to ensure sustainable development of the sector. This paper outlines the activities of the NGOs in the Asian region, especially those of the AFS and FHS/AFS in promoting aquaculture health management in Asia-Pacific. It also briefly examines the relationships between NGOs and the government agencies responsible for fish health management, and recommends ways by which NGOs may function more effectively

INTRODUCTION

Non-governmental organizations or NGOs, in a broad sense, embrace everything outside the public and private commercial sectors (Farrington *et al.*, 1993). However this definition is too broad and a more useful definition would be that NGOs are self-governing, independently financed organizations and societies that assign themselves specific roles, issues or disciplines paralleling governmental or intergovernmental organizations that have similar mandates.

NGOs have played a crucial role in various aspects of development in many countries around the world. This role is more significant in developing countries, where there is often a lack of political will for action plans to tackle important issues that require attention. NGOs also complement government approaches towards undertaking work on policy issues. Some NGOs have actively prioritized research according to the needs of the countries. In addition, NGOs have played an important role in networking and information sharing. Many operational NGOs work closely with grassroots groups via extension agents who live in rural communities, speak the local languages and share local concerns. NGOs have successfully operated by having the right expertise in the form of advisors or operation managers who plan, execute and implement without favor or prejudice.

The Agenda 21 (adopted by the UN Conference on Environment and Development on 14 June, 1992) program of action to be implemented from now and into the 21st Century, has stressed the crucial role of major social groups, including the NGOs, in all program areas.

This paper will discuss the role of NGOs in fish health management in the Asia-Pacific Region.

NGOs INVOLVED IN FISH HEALTH

Since fish health science is a relatively new field which has been neglected in the past and a field that is highly specialized, it is indisputable that the major organizations involved with fish health are scientific organizations, such as fisheries and veterinary societies. In the Asia-Pacific, the involvement of the fisheries societies has a longer history than that of the veterinary societies. Since the Asian Fisheries Society (AFS) has made the major contribution towards the science of fish health management, I would like to provide a brief background on the society.

The AFS is a scientific society formed in 1984 for fisheries professionals in Asia. The Society undertakes relevant and service-oriented activities which include:

- organizing regular triennial fora where fisheries professionals can interact and discuss broad issues and specific topics related to the region;
- publishing a scientific journal of an international standard;
- providing financial and technical support to young scientists from developing countries to conduct quality research; and
- addressing important fisheries issues via workshops and publications.

The Society has three branches; the Indian, Japanese and the Taiwan Branch. In addition, the China Fisheries Society and the Malaysian Fisheries Society are affiliated to the AFS. The difference between the affiliated societies and the branches of AFS is that the latter use the same constitution as the parent society while the former have their own constitutions.

ROLE OF AFS IN FISH HEALTH

Workshops and Fora

The AFS's role in fish health management began in 1986, when a one-day workshop on two topics, current fish disease problems in the region and the status of programs for fish quarantine, was held in Manila on 30 May, 1986. Twelve papers from seven countries were presented during the workshop. During the same period, many fish health scientists also presented scientific papers on fish health management issues at the First Asian Fisheries Forum, held in Manila during 26-31 May, 1986. Scientific papers related to fish health management were also presented during the subsequent triennial fora which were held in Japan (1989), Singapore (1992) and Beijing (1995).

Publication of the Journal

The Society began publishing *Asian Fisheries Science*, an international fisheries journal, in December 1987. To accommodate the growing number of manuscripts submitted, the journal increased the number of issues from the initial two in 1987 to three issues in 1990 and to four in 1994. The publication is receiving support from SEAFDEC, ICLARM and private companies such as Charoen Pokphand Group, Thailand, and the Hanaqua International Corporation and Fwu Sow Grains Products Co. Ltd from Taiwan. Many contributions on fish health have appeared in this journal. Special publications on specific topics or proceedings of workshops organized by the AFS are also published. To date, there are 12 publications of workshop proceedings and the first of this series focuses on fish health: *Fish Quarantine and Fish Diseases in South and Southeast Asia: 1986 Update* (Arthur, 1987).

Research Fellowship Awards

The AFS also provides research fellowship awards to promising young scientists in developing countries in Asia. The project is officially known as the Asian Fisheries Society Fellowship Award Scheme. The main target applicants are new Ph.D. and M.Sc. graduates.

The aim of this project is to:

- increase the technical competence of young scientists through the conduct of well-framed research which will produce useful results;
- develop a core of quality fisheries researchers whose outputs will enhance the development of fisheries science in the region and hence upgrade national and regional technical capabilities in Asia; and
- encourage and enhance linkages among fisheries scientists in the region and promote relationships between senior and young fisheries scientists.

This Fellowship Award Scheme was launched in 1987 with financial support from the International Development Research Centre (IDRC) of Canada. The Society provides technical and administrative support, with a number of its members serving as scientific advisors and project reviewers. Under this program US$ 10 000 to 12 000 is awarded to the scientist and matching funds for the project are given by the International Foundation for Science (IFS). Of the 38 research grants so far awarded, five were for fish health projects.

Training Attachment

More recently, the fellowship award scheme has been broadened to provide training to scientists from developing countries in more developed countries or in advanced institutions in the region. The two institutes that have signed a memorandum of understanding to provide such training are the SEAFDEC Aquaculture Department in Iloilo and the Taiwan Fisheries Research Institute (TFRI) in Keelung. One candidate from India is already undergoing training at the TFRI and another from Vietnam will be sent to SEAFDEC.

AFS FISH HEALTH SECTION

In January 1989, the formation of the Fish Health Section (FHS) of the AFS provided another avenue to focus on fish health issues. The purpose of the Section is:

- to promote effective interaction among persons involved in fish health research;
- to encourage and promote investigation and advancement in the knowledge of fish health;
- to focus attention on fish health problems by disseminating technical and other information; and
- to promote the proper implementation of effective fish health protection practices in the region.

The first meeting organized by the FHS was held in Bali in 1990 and was attended by 177 participants from 19 countries. During the four days, on 26-29 November, 77 oral and 18 poster presentations were made on various aspects of fish health. In conjunction, a special session on the use of chemotherapeutic agents in Asian aquaculture was also held. At the special session, papers were presented by participants from Australia, India, Indonesia, Iran, Japan, Malaysia, Philippines, Sri Lanka, Taiwan and Thailand. The proceedings of the first meeting were published as *Diseases in Asian Aquaculture I*, edited by Shariff, Subasinghe and Arthur (Shariff *et al.*, 1992).

The second symposium, convened in Phuket, Thailand, from 25-29 October, 1993, was attended by 170 participants from over 30 countries. Along with various fish health issues, shrimp health management was given special emphasis. There were 70 oral presentations and 36 posters on prevention of diseases in fish, shrimp, and other shellfish. These papers will be published shortly in *Diseases in Asian Aquaculture II* (Shariff *et al.*, 1995). Besides the formal presentations, the FHS also prioritized regional needs for fish health research. This exercise was held at the request of Strategy for International Fisheries Research (SIFR). The *Priorities for Disease Research: Towards Minimizing Losses from Diseases in Asian Aquaculture* was published in FHS Newsletter 4 (2), 1993.

The Third Symposium of the FHS will be held jointly with the World Aquaculture Society in Bangkok from 29 January to 2 February, 1996. This will be the FHS's first joint activity with another NGO from the West.

Besides organizing the fora and the special symposia, the FHS also publishes a newsletter which is well received by its members. Two issues are published annually and it features new research findings, updates on members, a list of recent publications and information on fish health activities in the region.

The FHS also publishes a series of regional bibliographies on all fish health research. Two such bibliographies, one covering Southeast Asia and another for Japan, have been published (Arthur, 1992, Wakabayashi, 1994) The former provides abstracts of all available literature covering Southeast Asia and includes English language abstracts, covering many articles published in national languages. The bibliography for Japan, a contribution of the Japanese Society of Fish Pathology, provides the titles of all the fish health publications in Japan. It contains 2 271 references which are listed under seven sections: cell lines, non-infectious diseases, viral, bacterial, fungal, protozoan, metazoan diseases and immunology. A bibliography on South Asia, which will include Bangladesh, India, Pakistan and Sri Lanka, is in preparation. The FHS has also proposed that all references cited in these bibliographies be housed at the FHS Resource Center at the Universiti Pertanian Malaysia. Currently the Resource Center has copies of all the articles cited in the Southeast Asian bibliography. The references for South Asia are already in UPM, but are not yet catalogued.

BRANCHES AND AFFILIATED SOCIETIES

Besides the parent society, the branches and the affiliated societies have also been involved in addressing fish health issues.

AFS Taiwan Branch

The AFS Taiwan Branch (formerly known as the Taipei Branch) was formed in 1985. It works closely with the Taiwan Fisheries Society in organizing fora and is one of the major contributors to the parent society by generating funds from the National Science Council, Council of Agriculture and from private companies. It publishes its own scientific journal which has featured several fish health papers. In 1994, the Taiwan Branch held an *International Symposium on Biotechnology Application in Aquaculture* from 5 to 10 December. Besides reproductive physiology and endocrinology and gene transfer, the other main topics for the meeting were disease diagnosis and therapy (see Kuo *et al.*, 1995).

AFS Indian Branch

The AFS Indian Branch was formed in 1986. Fish health issues have been addressed through annual fora and also in the regular quarterly newsletter. No special meetings to focus on fish health have been held but a substantial number of papers on the subject have been presented at annual meetings.

Taiwan Fisheries Society

The Taiwan Fisheries Society was established in 1955 and is one of the few societies in Taiwan to receive an annual grant of 400 000 NT from the National Science Council. The Society holds annual meetings where scientific papers are presented. The Society also publishes its own journal which has four issues a year and contains fish health papers.

Malaysian Fisheries Society

The Malaysian Fisheries Society (MFS), which was formed in 1987, has held annual seminars which include fish health presentations along with other scientific papers on fisheries. Most of these seminars were organized jointly with the Department of Fisheries, Malaysia and the Faculty of Fisheries and Marine Science, Universiti Pertanian Malaysia. On three occasions the seminars were sponsored by the Ministry of Science and Technology and, on two occasions, by the Fisheries Development Authority. In 1993, the MFS held a seminar specifically to address fish health issues. *Diseases in Aquaculture: the Current Issues* was co-sponsored by a private company which sponsored speakers from Thailand and Taiwan. This seminar was attended by 143 participants (see Subasinghe and Shariff, 1994).

OTHER NGOs

Japanese Society of Fish Pathology

The Japanese Society of Fish Pathology (until 1979 known as the Japan Research Group of Fish Pathology) was established in 1966. Its objectives are promotion of studies on fish pathology and dissemination of knowledge about fish diseases and related subjects. The Society publishes the scientific journal *Fish Pathology* and holds two meetings annually (spring and autumn) where more than 50 scientific papers are presented. Last year the Society started giving a one-week course on fish disease to entrepreneurs from the private sector.

The Veterinary Association of Malaysia

More recently, veterinary associations have also started to address fish health issues in their meetings. In November, 1994, the Veterinary Association of Malaysia, together with several other organizations and the Department of Veterinary Services, Malaysia, jointly held an *International Congress on Quality Veterinary Services for the 21st Century*. Fourteen papers on aquatic animal health were presented at this meeting.

CONCLUSIONS

In Asia, fish health management issues have been actively addressed by the fisheries societies and more recently and to a lesser extent, by the veterinary associations. Although there are other societies and organizations that have been indirectly involved in fish health issues, their activities have been restricted to publication of scientific articles in journals and have been small in number. These NGOs do not have a commitment to address fish health issues *per se*.

In organizing their meetings, the NGOs usually adopt a flexible approach, and they are free to choose the venue and the agenda, and to involve the expertise or invite the speakers of their choice. Several of the fora organized by the societies have been jointly held with the various departments related to fisheries and attended by officers from the government, university staff and members of the private sector. These fora are the only gatherings where there is a free flow of information between the various sectors participating. They thus enhance interaction between fisheries personnel from the private and the public sectors and facilitate transfer of technology from research institutions to the industry. The NGOs, with their pool of expertise, have generated several reports which serve as valuable references for policy makers.

However, in spite of their many advantages and achievements, NGOs face many constraints. One major constraint is the lack of financial resources that hamper activities. The government sector has considerable resources which are beyond the scope of the NGOs *e.g.*, well equipped research institutes and universities. Another dilemma is that the NGOs are often regarded with suspicion and as potential competitors by governments and other organizations. These constraints, however, should not be considered major drawbacks, since many members of NGOs are renowned experts in their fields and well respected in the scientific community and by policy makers. Thus, as long as there is a clear understanding as to the roles of the government sector and the NGOs, NGOs will always be seen as significant contributors to the development of the fisheries industry. However, it is important that the NGOs, after identifying the role they can play effectively, continue to develop linkages with the appropriate agencies. The NGOs role should be to provide the missing links or even to complement the ongoing process by strengthening the aspirations of the identified partners. In the process, it is also important that professionalism be maintained in all dealings. Future strategies of NGOs should be to communicate more effectively their visions and policies to the parties concerned, including governments and donor agencies, to ensure effective implementation. The NGOs are a foundation for good organizational operations and, as seen in the case of the AFS, an excellent form of networking which should be further utilized by the agencies concerned. Meanwhile, the establishment and maintenance of good relationships between all partners and the NGOs are crucial to their success.

REFERENCES

Arthur, J.R. (ed.), 1987. Fish Quarantine and Fish Diseases in South and Southeast Asia: 1986 Update. Report of the Asian Fish Health Network Workshop held in Manila, The Philippines, 30 May 1986. Asian Fish. Soc. Spec. Publ. No. 1, 86 p.

Arthur, J.R. (ed.) 1992. Asian Fish Health Bibliography and Abstracts I: Southeast Asia. Fish Health Sect., Asian Fish. Soc., Spec. Publ. No. 1, 252 p.

Kuo, C.-M., J.-L. Wu and P.-P. Hwang, 1995. Proceedings of the International Symposium on Biotechnology Applications in Aquaculture. 5-10 December, 1994, Taipei, Taiwan, ROC, Asian Fish. Soc. Spec. Publ. No. 10, 252 p.

Farrington, J., D.J. Lewis, S. Satish and A. Miclat-Teves, 1993. Non-governmental Organizations and the State in Asia: Rethinking Roles in Sustainable Agricultural Development. Routledge, London, 366 p.

Shariff, M., J.R. Arthur and R.P. Subasinghe (eds.), 1995. Diseases in Asian Aquaculture II. Proceedings of the Second Symposium on Diseases in Asian Aquaculture, 24-29 October, 1993, Phuket, Thailand. Asian Fish. Soc., Fish Health Sect., Manila, Philippines. 550 p.

Shariff, M., R.P. Subasinghe and J.R. Arthur (eds.), 1992. Diseases in Asian Aquaculture I. Proceedings of the First Symposium on Diseases in Asian Aquaculture, 26-29 November, 1990, Bali, Indonesia. Fish Health Sect., Asian Fish. Soc., Manila, Philippines, 587 p.

Subasinghe, R.P., and M. Shariff (eds.), 1994. Diseases in Aquaculture: the Current Issues. Proceedings of a Seminar Organized by the Malaysian Fisheries Society and the Department of Fisheries, Malaysia, 6 February, 1996, Kuala Lumpur. Malaysian Fish. Soc. Publ. No. 8, 173 p.

Wakabayashi, H. (ed.), 1994. Asian Fish Health Bibliography III. Japan. Asian Fish. Soc., Fish Health Sect. Spec. Publ. No. 3, 136 p.

Training and Extension in Aquaculture Health Management in the Asia-Pacific: Present Status and Future Requirements

Kamonporn Tonguthai

Aquatic Animal Health Research Institute
Directorate General for Fisheries
Kasetsart University Campus
Jatujak, Bangkok 10900
Thailand

Tonguthai, K. 1996. Training and extension in aquaculture health management in the Asia-Pacific: present status and future requirements. *In* Health Management in Asian Aquaculture. Proceedings of the Regional Expert Consultation on Aquaculture Health Management in Asia and the Pacific. R.P. Subasinghe, J.R. Arthur & M. Shariff (eds.), p. 60-74. FAO Fisheries Technical Paper No. 360, Rome, FAO. 142 p.

Abstract

Training and extension are an important components of a strategic program for health management in aquaculture. Considering the huge contribution made by Asia to the global aquaculture production and the importance of continuing production to meet future demands, Asian fish health personnel must be adequately trained in order to provide effective extension services to aquaculturists. Effective extension of sound fish health practices to the farmers themselves is also essential, so that they will be trained to manage their operations using methods which will reduce the likelihood of outbreaks of disease. This paper summarizes the present status of fish health training and extension in the Asia-Pacific and makes recommendations for the future.

INTRODUCTION

It is undoubtedly true that total world fisheries production from aquaculture is increasing. Approximately 84% of world aquaculture production comes from Asia (Nash and Kensler, 1990). Aquaculture is expanding along the lines of more intensive culture practices, with disease problems appearing to coincide with these practices. At present, the economic importance of aquatic animal disease control and health management is well recognized.

To develop better culture practices and reduce problems due to disease, advanced technology should be introduced. The techniques developed for dealing with cultured species in one country can be applied to similar species in other countries, especially in the same region. The most effective means of technology transfer is through training.

The governments of various countries in the Asia-Pacific Region realize the need to develop technology and to increase knowledge of aquatic animal health to achieve sustainable aquaculture. More institutes are now capable of conducting advanced research, and the number of experts in the various fields of fish disease is increasing. As a result of this progress, the present situation provides great possibilities for organizing various types of training within the region. This may lead to better cooperation among the countries.

TRAINING NEEDS IN AQUACULTURE HEALTH MANAGEMENT IN THE ASIA-PACIFIC REGION

Information on training and extension presented in this paper is based on the four following sources:

- ADB/NACA. 1991. Fish Health Management in the Asia-Pacific Region. Report on a Regional Study and Workshop on Fish Disease and Fish Health Management. ADB Agricult. Dep. Rep. Ser. No 1. Network of Aquaculture Centres in the Asia-Pacific Region, Bangkok.
- Anon. 1993. Priorities for disease research: towards minimizing losses from diseases in Asian Aquaculture. Asian Fish. Soc., Fish Health Sect. Newsletter 4(2): 4-5, 10 p.
- Tonguthai, K., S. Chinabut and I. H. Macrae. 1995. Regional institutional survey, April-May 1995. South East Asia Aquatic Disease Control Project, Aquatic Animal Health Research Institute (AAHRI), Bangkok, 10 p.
- U Khin Maung Aye and P. B. Bueno. 1994. An overview of fisheries development in Myanmar with special reference to opportunities for TCDC assistance, 3-7 December 1994. TCDC, Network of Aquaculture Centers in the Asia-Pacific Region, 12 p.

The present situation indicates that the services of fisheries experts have increased in the institutes in the Asia-Pacific Region. Experts offer their services mainly on a short-term basis. In addition, "on the job" or group training sessions are also organized depending on the experience of the participants. Among the 23 countries which participated in the meeting in Myanmar in December, 1994, at least six can offer services in health management (about 12 institutes) (see Table 1)

The ADB/NACA report (ADB/NACA, 1991) indicates that the level of capability in fish health management in the Asia-Pacific Region is extremely inadequate but also quite variable. The development of this expertise relies on the training of staff and the provision of facilities and must take into account the individual problems within each country. Virology has been identified as a new field for most laboratories in developing countries in the Asia-Pacific Region. At present, where shrimps have become a major species for culture, viruses appear to be one of the major pathogens causing shrimp disease. Knowledge concerning shrimp viruses, therefore, has developed dramatically in some institutes.

The number of institutes in the Asia-Pacific Region involved in fish disease control and fish health management in 1990 is given in Appendix I of ADB/NACA (1991). Institutes are also categorized according to level of expertise. However, if this is compared with the regional survey of the AADCP project conducted in 1995 (Tonguthai *et al*., 1995), some institutes can now be upgraded, for example, the Fisheries Research Institute (FRI) and Bangladesh Agricultural University (BAU) in Bangladesh; and the Faculty of Fisheries, Kasetsart University, Thailand. Progress in the upgrading of facilities can also be seen in other institutes, for example, the Brackishwater Aquaculture Development Centre (Jepara), the Research Institute for Freshwater Fisheries (Bogor) and the Fisheries Research Institute No. 1 (Hanoi). Expertise will not improve unless the training of staff is developed. Training and extension, therefore, are still in great demand. Even where expert services are offered by various institutes, the expertise seems to be limited to particular fields. The number of experts from each institute is also limited and may not be sufficient to provide service to their own countries.

Table 2 provides a summary of the needs for assistance in aquatic animal health management, and is based on information obtained from the meeting in Myanmar in 1994 (U Khin Maung Aye and Bueno, 1994).

The ADB/NACA report (ADB/NACA, 1991; see Appendix I) on fish health management institutes in the Asia-Pacific Region involved in research, diagnosis and extension appears to be comprehensive. It indicates that 23 of the 43 institutes involved in fisheries are concerned with fish disease control. In some, the level of expertise is very low. Therefore, there is a high demand for assistance in this field.

FUTURE REQUIREMENTS FOR TRAINING IN AQUACULTURE HEALTH MANAGEMENT

In summary, the greatest need for assistance by the 23 countries is in the field of shrimp diseases. It is clear that most future training should be in this field. However, basic training in fish disease diagnosis is still needed in some countries.

In intensive shrimp culture practice, management seems to play a major role in the disease problem. Knowledge of the diseases themselves may not always prevent the occurrence of the disease. Shrimp health management should be included in the training program.

There are quite a few countries in the Asia-Pacific Region where intensive shrimp culture is not fully developed and fish culture is more intensive. Shrimp health management, therefore, is not required in the near future in these countries.

The results of the regional survey carried out by the AADCP (Tonguthai *et al.*, 1995) in eight participating countries (Laos, Vietnam, Cambodia, Philippines, Malaysia, Indonesia, Bangladesh and Thailand) indicate that there have been improvements in laboratory facilities in all eight countries (but not all institutes) during the past four to five years. The other countries in the Asia-Pacific Region (except Laos and Cambodia) probably show similar trends. However, the number of trained staff is still limited, and there is, therefore, an urgent need to upgrade the existing staff who are actively involved in aquatic animal health. Training within the region is cost effective. Because fish disease problems are similar in many countries, the knowledge gained from training can easily be transferred from one country to another. In some specific fields where expertise is lacking in the region, it may be possible to make use of outside experts.

In any developing country, training organized at an international level relies mainly on donor funding. Without this support, the chances of setting up training programs are very slim, so donor agencies must be made to recognize the economic importance of fish health management to aquaculture sustainability, and funding should be allocated for training in this field.

Academic training programs are also needed. Unfortunately there are very few institutes offering training in fish health at a degree level. To my knowledge only one university in Southeast Asia, the Universiti Pertanian Malaysia (Faculty of Fisheries and Marine Science) offers post-graduate training in tropical fish health. There are quite a few universities in the Asia-Pacific Region with Faculties of Fisheries whose courses include aquaculture and fish disease specialities. It is necessary to encourage these universities to establish training programs in fish health or to include advanced courses in fish disease control and health management in their programs.

REFERENCES

ADB/NACA, 1991. Fish Health Management in the Asia-Pacific Region. Report on a Regional Study and Workshop on Fish Disease and Fish Health Management. ADB Agricult. Dep. Rep. Ser. No 1. Network of Aquaculture Centres in the Asia-Pacific Region, Bangkok.

Nash, C.E., and C.B. Kensler, 1990. A global overview of aquaculture production in 1987. World Aquacult. 21: 104-112.

Tonguthai, K., S. Chinabut and I.H. Macrae, 1995. Regional institutional survey, April-May 1995. South East Asia Aquatic Disease Control Project, Aquatic Animal Health Research Institute (AAHRI), Bangkok, 10 p.

U Khin Maung Aye and P.B. Bueno. 1994. An overview of fisheries development in Myanmar with special reference to opportunities for TCDC assistance, 3-7 December 1994. TCDC, Network of Aquaculture Centers in the Asia-Pacific Region, 12 p.

Table 1. Expert services in aquatic animal health offered to other countries.

Country	Institute	Field	Type of service offered[1]			
			OJ	G	ST	Dg
India	Central Institute of Freshwater Aquaculture (CIFA), Bhubaneshwar	Fish diseases	x	x	x	
	Central Institute of Brackishwater Aquaculture (CIBA), Madras	Shrimp diseases	x			
Indonesia	Research Institute for Freshwater Fisheries (RIFF), Bogor				x	
Malaysia	Faculty of Fisheries and Marine Science (UPM), Kuala Lumpur	Tropical fish health			x	x
Philippines	Southeast Asian Fisheries Development Centre (SEAFDEC), Iloilo, Philippines	Fish disease diagnosis	x	x	x	
Thailand	Aquatic Animal Health Research Institute (AAHRI), Bangkok	Fish and shrimp health management	x	x	x	
	National Institute of Coastal Aquaculture. (NICA), Songkhla	"	x	x	x	

[1]OJ = on-the-job training, G = group training, ST = short-term visits by experts, Dg = Degree level academic training.

Table 2. The need for assistance in aquatic animal health management by countries in the Asia-Pacific Region.

Country	Institute	Field	Type of service required[1]			
			OJ	G	ST	Dg
Bangladesh	Fisheries Research Institute (FRI), Myensingh	Fish and shrimp health management, especially bacteria and viruses	x		x	x
Cambodia	Department of Fisheries, Phnom Penh	Basic fish disease diagnosis	x			
People's Republic of China	Chinese Academy of Fishery Science (CAFS) (18 sub-institutes)	Fish and shrimp diseases	x		x	
	Freshwater Fisheries Research Centre (FFRC), Wuxi	Fish diseases	x			
Fiji	Fisheries Department (MAFF), Suva	Basic fish disease diagnosis	x			
India	Central Institute of Brackishwater Aquaculture (CIBA), Madras		x		x	

Table 2 (Continued.)

Country	Institute	Field	Type of service required[1]			
			OJ	G	ST	Dg
	Central Inland Capture Fisheries Research Institute (CICFRI), Barrackpore	Fish immunology			x	
	Central Institute for Fisheries Education (CIFE), Bombay	Fish and shrimp diseases	x	x	x	
	Central Institute of Freshwater Aquaculture (CIFA), Bhubaneswar	Shrimp diseases	x		x	
Indonesia	Brackishwater Aquaculture Development Centre (BADC), Jepara	Shrimp diseases	x		x	x
	Directorate General for Fisheries (DoF), Jakarta	Establishing a shrimp health management office			x	
Laos	Fisheries Division, Dept. of Livestock Vet. Science	Basic fish disease diagnosis	x			
Malaysia	Department of Fisheries, Kuala Lumpur	Establishing a National Fish Disease Centre			x	x
Myanmar	Department of Fisheries, Rangoon	Basic fish disease diagnosis	x		x	

Table 2 (Continued)

Country	Institute	Field	Type of service required[1]			
			OJ	G	ST	Dg
Nepal	National Fisheries Training Centre, Kathmandu	Fish disease	x			x
Papua New Guinea	Department of Fisheries and Marine Science	Fish diseases	x			
Philippines	Bureau of Fisheries and Aquatic Resources (BFAR), Quezon City	Shrimp diseases	x		x	x
Sri Lanka	National Aquatic Resource Agency (NARA), Colombo	Fish and shrimp diseases	x	x	x	
Thailand	Aquatic Animal Health Research Institute (AAHRI), Bangkok	Rapid diagnosis, immunology and vaccination			x x x	x
	National Institute of Coastal Aquaculture (NICA), Songkhla	Vaccination and immunity	x		x x	x
Vietnam	Fisheries Research Institute:	Fish and shrimp diseases				
	NO.1 Hanoi		x		x	x
	NO.2 Ho Chi Minh	"	x		x	
	NO.3 Natrang	"	x		x	

[1]OJ = on-the-job training, G = group training, ST = short-term visits by experts, Dg = Degree level academic training.

APPENDIX I

INSTITUTES IN ASIA INVOLVED IN RESEARCH, DIAGNOSIS OR EXTENSION ON FISH DISEASES AND FISH HEALTH MANAGEMENT (modified from ADB/NACA, 1991)

Country and Institute	Speciality[1]	Level[2]	Remarks
Bangladesh			
Fisheries Research Institute (FRI), Mymensingh	FF	*	
FRI, Chandpur	FF	*	
FRI, Khulna	C		Planning stage
FRI, Cox's Bazaar	C, MF		Planning stage
Bangladesh Agricultural University, Mymensingh	FF	*	Planned upgrading
China			
Yangtze River Fishery Research Institute	FF	**	
Freshwater Fisheries Research Centre, Wuxi	FF, M	**	
Zhejiang Institute of Fisheries, Hangzhou	FF	**	
Yellow Sea Fisheries Research Institute, Qingdao	C, SW	**	
Pearl River Fisheries Research Institute Guangzhou	FF	**	
South China Sea Fisheries Research Institute, Guangzhou	C, MF	**	
East China Seas Fisheries Research Institute Shanghai	C	*	
Institute of Hydrobiology, Wuhan	FF	***	
Shanghai Fisheries University, Shanghai	C, FF	**	

Appendix I (Continued)

Country and Institute	Speciality	Level	Remarks
Hong Kong			
Agriculture and Fisheries Department	FF, MF	**	
India			
Central Institute for Freshwater Aquaculture Bhubaneswar	FF	***	
Central Inland Capture Fisheries Research Institute, Barrackpur	FF	**	Planned upgrading
Central Institute for Brackishwater Aquaculture, Madras	C, MF		Planning stage
Central Marine Fisheries Research Institute, Bombay	FF, M	*	
Central Institute for Fisheries Education	FF		Planning stage
College of Fisheries, Mangalore	FF	**	Planned upgrading
Indonesia			
Freshwater Aquaculture Development Centre (FADC), Sukabumi	FF	*	FADC, BADC and NSDC are under the Directorate General of Fisheries; RIFF and RICA are under the Central Research Institute for Fisheries
Brackishwater Aquaculture Development Centre (BADC), Jepara	C	**	
National Seafarming Development Centre (NSDC), Lumpung	MF	*	
Research Institute for Freshwater Fisheries (RIFF), Bogor	FF	***	
Research Institute for Coastal Aquaculture (RICA), Maros	C	**	

Appendix I (Continued)

Country and Institute	Speciality	Level	Remarks
Research Institute for Veterinary Science, Bogor	FF	**	Collaborates with RIFF on virology
Bogor Agricultural University, Bogor	FF	**	
Japan			
National Research Institute of Aquaculture, Tamaki	FF, M, MF	*****	Major fish disease research institute under the Fisheries Agency
Korea (Republic of)			
National Fisheries Research and Development Agency	M, MF, SW	***	
National Fisheries University of Pusan	FF, M, MF	***	
Yosu Fisheries University	FF	***	Estimated level of expertise
Gunsan Fisheries University	FF	***	Estimated level of expertise
Malaysia			
Brackishwater Fisheries Research Centre, Johore	C	**	
Freshwater Fisheries Research Centre, Malacca	FF	**	
Fisheries Research Institute, Penang	M, MF	**	
Universiti Pertanian Malaysia, Serdang	C, FF	***	
Universiti Sains Malaysia, Penang	FF, MF	***	
Myanmar			
Fisheries Research Institute, Yangon			Planned to contain a fish health laboratory

Appendix I (Continued)

Country and Institute	Speciality	Level	Remarks
Nepal			
National Centre for Fish Health Management, Kathmandu	FF		Planned only
Fisheries Development Centre, Janakpur	FF	*	
Pakistan			
National Agricultural Research Centre, Islamabad	FF	*	
Fisheries Research and Training Institute, Lahore	FF	*	Planned upgrading
Philippines			
Bureau of Fisheries and Aquatic Resources Quezon City	C, FF, MF	**	
SEAFDEC AQD, Tigbauan	C, MF, SW	***	
SEAFDEC AQD, Binangonan	FF	**	
Marine Science Institute, University of Philippines	SW	**	
University of Philippines in the Visayas	C, FF	**	
Central Luzon State University	FF	**	

Appendix I (Continued)

Country and Institute	Speciality	Level	Remarks
Papua New Guinea			
Department of Agriculture and Livestock		*	Mainly animal diseases with little work in fish
Singapore			
Primary Production Department, Marine Aquaculture Section	C, MF	**	
Freshwater Fisheries Section	FF	**	
City Veterinary Laboratories	FF, MF	****	Reference Centre for cases from the marine and freshwater sections
Pig and Poultry Research and Training Institute	C, FF, MF	***	
National University of Singapore	FF	***	Planned activities in fish disease management
Sri Lanka			
Freshwater Fisheries Station, Dambulla	FF	*	
National Aquatic Resources Agency	C	**	
University of Colombo	FF	**	Not operational at the present time
Taiwan			
Council of Agriculture Regional Fish Disease Centres	C, FF, MF	****	
National Taiwan University	C, FF, MF	****	

Appendix I (Continued)

Country and Institute	Speciality	Level	Remarks
Thailand			
National Inland Fisheries Institute	FF	***	Upgrading in progress
National Institute for Coastal Aquaculture	C, MF	***	
Coastal Aquaculture Development Division	C, MF	**	
Satul Coastal Aquaculture Centre	MF	**	
Suphanburi Freshwater Fisheries Centre	FF	**	
Rayong Coastal Aquaculture Centre	M	**	
Khung Kraben Bay Royal Development Study Centre	C	**	
Chulalongkorn University Veterinary School	C, FF, MF	***	
Department of Fisheries, Kasetsart University	C, FF	**	
Vietnam			
Aquaculture Research Institute No. 1	FF	*	
Aquaculture Research Institute No. 2	C, FF	*	
Aquaculture Research Institute No. 3	MF	*	
Research Institute for Marine Products	C, MF	*	
Fisheries University, Nha Trang	FF	*	

[1]C = crustaceans (mainly penaeid shrimps), FF = freshwater finfishes, M = molluscs, MF = marine finfishes, SW = seaweeds.

[2]* - ** = major upgrading needed, *** = upgrading needed in specialist areas, ****-***** = good all-round facilities

Health Management Strategy for a Rapidly Developing Shrimp Industry: An Indian Perspective

C.V. Mohan

Department of Aquaculture, College of Fisheries,
Mangalore-575002, India

Mohan, C.V. 1996. Health management strategy for a rapidly developing shrimp industry: an Indian perspective. *In* Health Management in Asian Aquaculture. Proceedings of the Regional Expert Consultation on Aquaculture Health Management in Asia and the Pacific. R.P. Subasinghe, J.R. Arthur & M. Shariff (eds.), p. 75-87. FAO Fisheries Technical Paper No. 360, Rome, FAO. 142 p.

Abstract

Over the last five years, the shrimp farming industry of India has transformed from a traditional shrimp trapping system to a capital oriented semi-intensive system. Availability of large areas of suitable coastal land and the recent liberalization policy of the government encouraged investment, leading to a phenomenal growth of this industry. More than 80 000 ha of coastal land have been brought under culture, producing about 75 000 mt of shrimp in 1994. The 100% export-oriented shrimp farming industry has generated rural employment and the development of integrated industries. Since July 1994, the industry has been under the grip of disease, two epizootics of viral etiology having nearly wiped out several farms along the east coast. In spite of observing "crop holiday" and following standard disinfection procedures, the disease has reappeared in farms where trial restocking has been attempted. Ecological degradation, intensification, the import of live seed, bad water management practices, high stocking densities, natural calamities etc. are among the causes proposed for the current catastrophe. Socio-economic implications, mangrove destruction and salination of agricultural land and freshwater bodies are some of the negative impacts of the industry's unregulated and hasty expansion. India has a vast potential for shrimp farming, and the industry has enormous potential to generate much-needed foreign exchange, rural employment and the development of ancillary industries. However, to achieve sustainability, it must develop in a planned and regulated manner. Hence, effective health management strategies, guidelines and regulations are urgently needed. This paper reviews the current status of the industry, examines current disease problems and proposes practical long-term health management strategies and regulations to develop a sustainable industry.

INTRODUCTION

Traditional shrimp trapping systems have been in practice in the low-lying brackishwater areas of Kerala and West Bengal for several years. These systems are largely seasonal in nature. Autostocking aided by tides is practiced, and managerial inputs are minimal. Production from these traditional trapping systems is very low, averaging 100-500 kg/ha/crop.

India has vast potential for the development of commercial shrimp farming, being blessed with 1.2 million ha of coastal brackishwater land. Many coastal areas are rich in natural seed resources, and broodstock collection grounds are said to be ideal for commercial exploitation. The industry is basically export-oriented and has enormous potential to generate much needed foreign exchange. Short development periods, high return on investment and good international demand are positive features which should encourage the development of the industry. In addition, the industry can create significant employment in rural areas, including employment for women.

Realizing this enormous potential, the Government has identified aquaculture as a "thrust area" under "extreme focus" for augmenting exports and earning much-needed foreign exchange. To encourage shrimp farming ventures, the Government has recently implemented a series of liberalization policies. Restrictions on the importation of aquaculture machinery, feed and aqua-chemicals have been eased, and import duties on specific items like feed have been completely slashed. Steps have been taken to encourage foreign equity participation in joint ventures and full repatriation of profits by foreign collaborators has been allowed. Five-year tax exemption has been given to companies which are 100% export oriented.

PRESENT STATUS OF THE INDUSTRY

Realizing the enormous potential for investment opportunities and taking advantage of the government liberalization policy, several national and international corporate houses, private companies and individual entrepreneurs have embarked on setting-up shrimp hatcheries and farms. Nearly 80 000 ha is under extensive or semi-intensive culture, and the quantity of farmed shrimp produced in 1994 was around 75 000 metric tonnes (mt). The area being brought under culture is increasing rapidly. At present India is ranked fourth in the world in cultured shrimp production.

In the last two to three years, growth in the shrimp hatchery sector has been phenomenal. Before 1993, seed availability was a major bottleneck. With the establishment of 100 state-of-the-art hatcheries with a total capacity of 4 billion seed per annum, this problem seems to be temporarily solved. Availability of good quality feed has not been a serious problem. Most feed is currently imported. Realizing the potential demand that exists for shrimp feed, several Indian and foreign joint-venture companies have established feed mills. Aquaculture machinery and aquachemicals are being regularly imported.

Resource mobilization and financing for the development of shrimp farming have been relatively well organized, because shrimp farming is no longer seen as a traditional occupation but as a capital intensive industry. Public and private banks are extending loans to the industry on a priority basis. Several private companies have gone public and have raised sufficient funds through public issue of equity shares. Insurance companies are also involved, with risks associated with natural calamities and diseases being insured and crop losses protected.

The concept of satellite and franchise farming as promoted by big industrial houses is well established and is working reasonably well. In these systems, the industrial house acts as the nucleus for several hundred small- and medium-scale extensive farmers. Technical input in the form of consultancy and other inputs in the form of seed, feed, and disease and water quality management are provided. The end product is bought back by the company at the prevailing market price. In addition to satellite farming, many small companies have long-term buy back arrangements with big industrial houses and multinational companies.

The booming shrimp industry has encouraged the development of several small-scale aqua-related ancillary industries in rural areas and has created many employment opportunities for the rural poor. In the last three to five years, the industry has grown by leaps and bounds, but still only a small percentage of the available 1.2 million ha of suitable coastal land has been brought under culture. In this short period, the industry has come face to face with many crippling problems. The repercussions of unregulated expansion are already being felt in many quarters. The prospect of rapid development and fast returns has led to haphazard and unorganized growth and the "fast buck syndrome" has overshadowed all relevant scientific and ecological considerations. Unplanned and unregulated mushrooming of farms around available water sources has led to negative impacts, such as serious ecological degradation, socio-economic problems, social tensions, mangrove destruction, isolation of villages and salination of agricultural land and freshwater wells.

The expansion of the industry and the desire of the farmers to make quick money have encouraged pseudoconsultants and quacks. The easy availability of a large quantity of inferior quality chemicals has led to an indiscriminate use of antibacterials and other chemicals. Bacterial drug resistance (Karunasagar *et al.*, 1994), tissue residues etc. are some of the serious problems which need to be addressed.

Aquaculture waste management has become a very serious issue. The majority of the farms use source water as a dumping ground for waste discharge. Raw effluents rich in organic matter and waste feed are released directly into water sources without any treatment or settlement. There are no community joint agreements on coordinated arrangements for water intake and effluent discharge. This has led to problems connected with disease transmission and bad water quality, and little attention is being given to pond drying, disinfection and waste removal between crops.

Crop losses due to natural calamities like flooding and cyclones have become common. Absence of buffer zones and destruction of mangroves have further aggravated these problems. Since July 1994, two viral diseases (yellowhead disease (YHD), and possibly systemic ectodermal and mesodermal baculovirus (SEMBV)) have had disastrous effects on the booming shrimp industry, causing losses estimated at around Rs 600 crores (US$ 17.6 million). Since January 1995, a fallow period, or "crop holiday," has been observed in many farms along the east coast. In spite of observing "crop holiday" and adopting standard disinfection programs, the disease has reappeared in many of the farms where trial restocking is done. It appears that the industry will need some time before it can regain its original tempo and status.

In view of the current industry crisis, there is an urgent need to develop and execute a rational health management strategy and to frame suitable legislation and guidelines for ensuring the development of sustainable shrimp farming. The shrimp farming industry in India needs to be developed on sustainable lines because there is still vast potential for growth, and significant potential for generating foreign exchange and rural employment.

MAJOR HEALTH PROBLEMS AND THEIR CAUSES

In hatcheries, luminescent vibriosis (*Vibrio harveyi*, *V. splendidus*, *V. parahaemolyticus*), external fouling (protozoan ciliates like *Zoothamnium* and *Vorticella*, filamentous bacteria like *Leucothrix*) and larval mycosis (*Lagenidium*) have been consistently recognized as major health problems (Felix *et al.*, 1994). In some hatcheries monodon baculovirus (MBV) has been identified as a problem (Felix and Devaraj, 1993; Ramasamy *et al.*, 1995).

The majority of the pathogens encountered in hatcheries are normal components of the marine ecosystem. Opportunistic vibrios and external foulers become a serious problem in badly managed hatcheries. Excess stress, bad water quality and inefficient disinfection programs are some of the causes of hatchery problems. With regard to MBV, it is well known that most broodstock are natural carriers. MBV becomes a serious problem only in stressed hosts and in hatcheries lacking organized and efficient disinfection programs.

Prior to April 1994, Indian shrimp farms did not experience any serious disease problems. Conditions such as external fouling, shell and appendage necrosis, black gills, luminescent vibriosis, systemic vibriosis, chronic vibriosis, muscle cramp etc.

were regularly reported from farms along the east and west coast. These health problems were responsible for low levels of mortality in several farms but not for mass mortalities over large areas.

Stressed shrimp tend to be sluggish and do not exhibit grooming behavior. Colonization by external foulers is very common in weak and unhealthy shrimp, and severely fouled shrimp are susceptible to opportunistic secondary invaders like vibrios. The majority of health problems in Indian shrimp farms can be traced to excess stress associated with intensification, poor water quality, pond bottom deterioration due to waste accumulation, benthic algal growth and subsequent decomposition, lack of reservoir tank facilities for carrying out suitable prophylaxsis and bad quality source water.

THE CURRENT CRISIS

Since July 1994 two diseases of suspected viral etiology (YHD and SEMBV?) have had disastrous effects on shrimp farming which has upset the industry (Shankar and Mohan, 1994; Mohan and Shankar, 1995). YHD first occurred in several farms located on either side of Kandaleru Creek, in Gudur, Andhra Pradesh, along the east coast of India. All farms in this region were receiving and discharging water into Kandaleru Creek; none had reservoirs or sedimentation tanks The disease affected only tiger shrimp (*Penaeus monodon*) (40-70 days old) in grow-out ponds. White shrimp (*P. vannamei*) stocked separately or together with tiger shrimp were not affected. Affected shrimp did not feed and had empty guts. At first, a large number of moribund shrimp were found near the pond margin and surface. The hepatopancreas of affected shrimp was reddish yellow or pale yellow in color and overall, the cephalothorax was yellow. Mass mortalities occurred within two to three days of the appearance of one or all of these clinical signs. To save the crop, many farmers resorted to emergency harvest.

In clinical signs, nature of mortality, species and size affected and, importantly, yellow coloration of the cephalothorax, the disease was very similar to yellowhead disease (YHD) caused by yellowhead baculovirus (YHBV). Confirmatory diagnostic reports are still not available.

The second disease, now recognized as white spot disease, affected the majority of farms in Andhra Pradesh and Tamil Nadu between November 1994 and January 1995. According to latest information, it is still spreading along the east coast to many parts of Orissa and West Bengal.

White spot disease affected all size groups of both tiger and white shrimp. Affected shrimp did not feed and had empty guts. Mortalities were low during the initial two to three days. A large number of circular white patches or spots were seen, first in the under surface of the cephalothorax and later in the carapace over the abdominal region. Mass moralities occurred within seven to 10 days of the first clinical signs.

In clinical signs, nature of mortality, and species and size affected, the disease is similar to IHHN caused by IHHNV. However, recently published work suggests that white spot disease is caused by systemic ectodermal and mesodermal baculovirus (SEMBV), a virus which has been responsible for the collapse of the industry in many parts of East and Southeast Asia (Nash, 1995).

It is estimated that these two diseases have caused a loss of nearly Rs 600 crores (US$ 17.6 million) to the industry. The Marine Products Export Development Authority (MPEDA), concerned research organizations and experts from the Asia-Pacific Region have all recommended "crop holiday," waste removal from ponds, pond drying and disinfection programs to prevent recurrence of the disease.

INSTITUTIONS AND THEIR ACTIVITIES

The shrimp farming industry has grown very fast and in a haphazard manner. This rapid growth is one of the reasons for the lack of guidelines, feasibility studies, research input etc. to the industry. Below is a brief account of the various institutions involved in the industry and their activities.

Marine Products Export Development Authority (MPEDA)

MPEDA has been instrumental in promoting shrimp culture in India, setting up model state-of-the-art hatcheries and demonstration farms in many parts of India, and has successfully demonstrated the techno-economic viability of commercial shrimp farming. In addition, MPEDA organizes annual INDAQUA shows, workshops, training programs, expert consultations etc., involving farmers, industry and experts from India and abroad. As a part of the promotional program, MPEDA is also extending subsidy facilities to shrimp farming and participating in the equity of public limited companies.

MPEDA is actively associated with the ministry in framing policy guidelines and regulations relating to aquaculture. It can take the role of a nodal agency to implement policy regulations in different shrimp farming states.

Brackishwater Fish Farmer's Development Agency (BFDA)

Many of the maritime states engaged in shrimp culture have established BFDAs at the district level. Normally officers from the respective state fisheries departments are assigned to look after the activities of these agencies. These agencies are involved in identifying and earmarking government land under their jurisdiction suitable for shrimp culture and are responsible for leasing out this land to unemployed youth, people from the poor sections of society, individual entrepreneurs and companies. Organization of a large number of promotional programs, workshops, training programs etc. is also undertaken. In many states, BFDAs have successfully established model hatcheries and farms to demonstrate their techno-economic viability. Subsidy for shrimp farming originating from the individual state governments is routed to the farmers through BFDAs.

Fisheries Colleges and Research Centers

Teaching and research institutions coming under the University of Agricultural Sciences of Karnataka, Tamil Nadu, Kerala, Andhra Pradesh, Maharastra, Orissa and Gujarat are very closely associated with the activities of the rapidly developing shrimp industry of India. Fisheries colleges have been very successful in providing much needed technical manpower to the industry. These institutes and colleges are currently engaged in research in areas like shrimp nutrition, water quality, disease etc. Extension and training programs, and workshops are also routinely undertaken.

Central Fisheries Research Institutes

Several central research institutes *i.e.*, Central Institute of Brackishwater Aquaculture (CIBA), Central Marine Fisheries Research Institute (CMFRI) and Central Institute of Fisheries Education (CIFE), are involved in research and extension activities and are closely associated with the shrimp industry. They also conduct workshops, training programs etc. for the benefit of farmers, technical personnel and the industry.

Non-Governmental Organizations

Several NGOs have been recently established. The Aquaculture Foundation of India (AFI), Karnataka Shrimp Farmer's Forum and the Karnataka Aquaculture Society are some of the leading organizations which are actively involved in promoting interaction between farmers, evolving joint agreements on water and waste management, and conducting training programs and workshops.

Big Industrial Houses

Many of the big industrial houses have established their own research and development (R&D) facilities. Some offer diagnostic services, organize workshops and training programs, publish extension literature etc.

Banks and Financial Institutions

Banks and financial institutions are heavily involved in extending loans for setting-up hatcheries, farms, feed mills, infrastructure, R&D laboratories etc.

Insurance Sector

Several insurance companies are involved in insuring against crop losses due to natural calamities and diseases.

International Organizations

International organizations like INFOFISH and the Asian Shrimp Culture Council (ASCC) disseminate a lot of useful information on feed, water quality, disease management etc. through newsletters. In addition, these organizations also conduct training programs and workshops involving experts from India and abroad for the benefit of the industry.

FUNDING SUPPORT

Until recently, funding support to shrimp disease research was very limited. The recent emergence of two deadly viral diseases and the trail of destruction that they have left behind have highlighted the need for systematic and coordinated research on shrimp diseases and their management. National and international funding agencies have now recognized the need to fund research so as to develop sustainable shrimp farming in India. Several research projects are in the pipeline and many more are likely to be funded in the near future. Funding support and coordination of research effort are needed within the country. The following are some of the national and international agencies which have funded or are likely to fund shrimp disease research in India:

- Indian Council of Agricultural Sciences (ICAR)
- Department of Biotechnology (DBT), New Delhi
- Department of Science and Technology (DST), New Delhi
- Department of Ocean Development (DOD), New Delhi
- National Bank of Agriculture and Rural Development (NABARD)
- Marine Products Export Development Agency (MPEDA)
- Various state fisheries departments
- Corporate houses
- International Foundation for Science, Sweden
- World Bank
- Australian Centre for International Agricultural Research (ACIAR)

ACHIEVEMENTS

Indian shrimp farming is still in its infancy. The research effort that has gone into shrimp disease studies has been very limited and hence very limited progress has been made towards health management. The majority of the health management techniques and chemotherapeutic practices that are being followed are largely adopted from Southeast Asian countries. The marketing strategy adopted by the international feed suppliers has helped in the acceptance and implementation of several health management strategies. Some of the noteworthy developments that have taken place in the last three to five years in the Indian shrimp industry are discussed below.

Recurrent problems associated with luminescent vibriosis, external foulers, larval mycosis and MBV are being effectively managed in many hatcheries. The current practice of taking good quality sea water from bore holes sunk in the sea bed near the coast has kept many problems at bay. The use of pressurized sand filters, UV irradiation systems, better disinfection programs, and the effective use of antibacterials and other chemotherapeutants have enabled the hatcheries to function successfully in recent years. Chemotherapy for secondary and systemic vibriosis and external foulers in grow-out farms has been successfully adopted and standardized.

Having realized the importance of health management for successful aquafarming, the Government of Karnataka has established a Disease Diagnosis Centre at the College of Fisheries, Mangalore. This center has a mandate to offer disease diagnostic and health management services to the farmers of the state. Several maritime states are considering establishing such centers.

Several research projects on shrimp diseases have been initiated. Some of the priority areas that have been identified are development of immunodiagnostics for screening and rapid diagnosis, development of vaccines and immunopotentiators, shrimp virology and evaluation of routine health monitoring programs. Over the last two years, a large number of training programs and workshops involving experts from India and abroad have been organized. These training programs and the current crisis the industry is facing has certainly impressed upon farmers the need for effective health management strategies and proper regulations for sustainable growth of the industry.

Many public limited companies have successfully established R&D centers. These centers will play a vital role in implementing rational health management practices over large areas through satellite and franchise farming.

In view of the current crisis, the Government of India has decided to issue aquaculture guidelines to all the concerned states to ensure the safety and sustainability of all aquafarming operations. Earmarking specific areas, statutory "crop holiday" or crop rotation, restricting maximum stocking density to < 30 m^2 and complete drying of the pond bottom for two to three months are some of the areas in which the Government has decided to issue guidelines.

CONSTRAINTS IN HEALTH MANAGEMENT

Management Constraints

Short gestation periods and high returns on investment have overshadowed many scientific and ecological considerations during the development of the shrimp industry. Adopting effective health management has therefore been very difficult.

- Dependence on a common water source is the root cause of many problems Most farms do not have reservoir and sedimentation tank facilities. Lack of reservoir tanks makes it impossible to undertake prophylactic measures, while lack of settlement tanks for sedimentation and disinfection of waste water before discharge adds to the problem of health management. The majority of the farms are owned by small farmers (< 5-10 ha) and it will be difficult for them to make provisions for reservoir and sedimentation tanks.
- Lack of a sufficient number of disease diagnostic centers.
- Lack of a sufficient number of trained personnel for disease diagnosis.
- Failure to implement routine health monitoring programs. From the point of view of chemotherapy, the normal practice of diagnosing the cause of a disease during a mortality serves little purpose.
- The nature of the farms and the dependence on common water source make it very difficult to follow health management strategies at the individual farm level.
- Lack of coordinated joint agreements between farms on waste disposal, effluent discharge and water intake.
- Lack of awareness among farmers on the need for proper disposal of dead shrimp and waste water during and after a disease outbreak.

Constraints in Regulation

The nature and extent of the shrimp farming industry in India make it very difficult to follow standard and scientific health management strategies. Lack of adequate regulations and absence of nodal agencies to implement such regulations are serious constraints to enforcing strict health management over a large area. At the national level, there is no comprehensive aquaculture policy and there is no nodal agency to enforce regulations. Some of the important issues which need immediate attention are:

- The process of disease notification is absent in the country and hence it is difficult to restrict in-country movement of infected seed and broodstock.
- Quarantine programs for new introductions and imports are not effectively implemented.
- Disease certification programs are not practiced.
- Rapid disease screening systems are not available.
- Enforcing 100% "crop holiday," if and when required, is very difficult.
- No effective regulations exist to check the misuse of antibacterials and other therapeutic chemicals.
- Entry of pseudoconsultants and quacks has led to large-scale availability of bad quality chemicals and their misuse.
- Lack of effective regulation and its implementation to halt the ecological degradation associated with unplanned mushrooming of farms around available water sources makes the problem of health management worse.

RECOMMENDATIONS

The Indian shrimp industry is still at a very early stage of development. The industry has enormous potential to generate much needed foreign exchange and significant rural employment. In the interest of the nation, it is essential that it develop in a planned and regulated manner on a sound ecological and scientific basis. This can be ensured only if proper future directives are developed and effectively implemented. The future directives should be in the areas of policy and regulations, management, and research and development.

Regulations

Practical and rational regulations should be developed and implemented to guide the development of the industry. Some important areas which need to be considered are:

- Development of a comprehensive national policy on aquaculture.
- Restriction of uncontrolled shipment and importation of broodstock and larvae.
- Implementation of organized quarantine, eradication and health certification procedures.
- Regulation of man-made canals and creeks bringing brackishwater deep into agricultural land.
- Conversion of agricultural land to shrimp ponds.
- Regulations to check the misuse of chemicals.

- Mandatory waste removal, drying and disinfection programs between crops at farms.
- Enforcement of complete "crop holiday," when required.
- Restriction of the maximum stocking density to < 30 m^2.

Management

Adopting some of the following management measures would go a long way in aiding the development of sustainable shrimp farming in India:

- Routine health monitoring should be mandatory for all farms. Early detection of disease would permit rapid treatment, adjustment of management or emergency harvest.
- Disease notification should be mandatory, so that restrictions can be enforced on movement of live seed and broodstock within the country.
- Routine prophylactic husbandry practices should be made compulsory to obtain licenses, bank financing, insurance coverage, subsidy etc.
- Each farm should allocate areas for reservoir and settlement tanks.
- Waste management should be strictly enforced (sedimentation tank).
- Organized disinfection programs should be implemented.
- Community agreements in water management (water intake and waste water discharge) should be developed.
- Closed/semiclosed systems of culture should be considered.
- Mangrove destruction should be prevented.

Research and Development

Research and developmental work in the following areas would help evolve standard scientific health management tools:

- Exploring the use of specific pathogen-free broodstock and healthy post-larvae.
- Exploring the development and use of virus-resistant species.
- Developing immunodiagnostic kits for rapid disease diagnosis, and screening of broodstock and post-larvae.
- Determining the carrying capacity of brackishwater sources before developing farms and earmarking areas for shrimp farming.
- Establishing disease diagnostic centers in shrimp farming belts.
- Creating a national center of excellence in shrimp disease research.
- Developing immunoprophylactics, including vaccines, anti-stress factors, immunostimulators etc.
- Initiating work on kinetics of antibacterials in shrimp tissue, tissue residues, bacterial drug resistance etc.

- Developing bioremediation technology for shrimp farming.

ACKNOWLEDGMENTS

The sponsorship received from FAO and the FHS for participating in the expert consultation is gratefully acknowledged. I thank the University of Agricultural Sciences for permitting my participation in this workshop.

REFERENCES

Felix, S., and M. Devaraj, 1993. Incidence of destructive MBV and IHHNV in a commercial shrimp hatchery. Sea Food Export J. 25:

Felix, S., I. Rachel, and V. Sunderraj, 1994. Wide prevalence of ciliate infestation in shrimp aquaculture systems of Tamil Nadu. Sea Food Export J. 25:

Karunasagar, I., R. Pai, G.R. Malathi, and I. Karunasagar, 1994. Mass mortality of *Penaeus monodon* larvae due to antibiotic resistant *Vibrio harveyi* infection. Aquaculture, 128: 203-209.

Mohan, C.V., and K.M. Shankar, 1995. Mass mortality of shrimps along the east coast of India - viral pathogen suspected. Fishing Chimes, (in press).

Nash, G., 1995. SEMBV - an emerging viral threat to cultured shrimp in Asia. Asian Shrimp News, 20:

Ramasamy, P., G.P. Brennan and R. Jayakumar, 1995. A record and prevalence of monodon baculovirus from postlarval *Penaeus monodon* in Madras, India. Aquaculture, 130: 129-135.

Shankar, K.M., and C.V. Mohan, 1994. Kandeleru fed brackishwater farms in Gudur, AP - viral pathogen suspected causing massive tiger shrimp mortality. Fishing Chimes, September, 1994.

A Review of Traditional and Innovative Aquaculture Health Management in the People's Republic of China

Jiang Yulin

Shenzhen Animal and Quarantine Bureau
40 Heping Road, Shenzhen 518010
People's Republic of China

Jiang Yulin. 1996. A review of traditional and innovative aquaculture health management in the People's Republic of China. *In* Health Management in Asian Aquaculture. Proceedings of the Regional Expert Consultation on Aquaculture Health Management in Asia and the Pacific. R.P. Subasinghe, J.R. Arthur & M. Shariff (eds.), p. 88-103. FAO Fisheries Technical Paper No. 360, Rome, FAO. 142 p.

Abstract

In 1994, China contributed 60.4% to the total world aquaculture production. Although freshwater and marine fish production increased rapidly over the past decade, marine shrimp culture suffered serious losses due to disease outbreaks. It has been said that these outbreaks of virus disease are partly management oriented. In order to find an effective solution to prevent disease outbreaks, the concept of "putting prevention first" has now been introduced, and this paper describes the current health management strategies that are being practiced in Chinese aquaculture.

AQUACULTURE IN THE PEOPLE'S REPUBLIC OF CHINA

Prior to 1980, fish production in China was dominated by marine capture fisheries and aquaculture accounted for less than one million metric tonnes (mt). This limited production, which was inadequate to meet demands, came mainly from the extensive freshwater pond culture of four species of Chinese carps; grass carp (*Ctenopharyngodon idella*), black carp (*Mylopharyngodon piceus*), silver carp (*Hypophthalmichthys molithrix*), and bighead carp (*Aristichthys nobilis*). The culture of finfish in lakes and reservoirs and of marine shellfish was also practiced, but their contributions were not significant. Since then, aquaculture has developed rapidly, with production increasing to some 8 million mt at present. Indeed, aquaculture has achieved such importance that by 1992 it contributed more than half of the total aquatic production.

Freshwater Aquaculture

Intensive high-production fish ponds for the culture of market fish have been set up in the suburbs of large cities, with a total area of 167 000 ha under culture in 1991 (11 000 ha in the suburbs of Shanghai alone), and yields reaching 11.4 mt/ha. Pond culture of fish has always been the mainstay of Chinese aquaculture and an important channel for increasing the output. For example, in 1988 some 1.4 million ha of ponds produced 2.9 million mt of fish *i.e.*, 35% of the culture area contributing 75% of the fish produced via freshwater culture. During the 10 years from 1978-1988, production increased five fold, while the area of ponds doubled.

There are lakes and reservoirs covering an area of 12 million ha, but about only one fifth of them are used for aquaculture. Low density culture is carried out in large lakes and intensive culture in cages and enclosures. Efficiency is generally low, the average output being 60 kg/ha for lakes and 210 kg/ha for reservoirs. Cage-culture production is on the order of 600 mt/ha.

Industrial fish farming, which includes the use of warm flowing water from power stations, rainbow trout (*Oncorhynchus mykiss*) culture using cold flowing water, and industrial fish culture using confined circulating water, has developed very quickly. In Shanghai, yields have reached 270 mt/ha for rainbow trout and 750-1300 mt/ha for tilapia. Many freshwater species are cultured in China, such as rainbow trout, eels, tilapias, shrimps, pearl shell, frogs, freshwater crabs, turtles etc. In Hubei Province, for example, freshwater aquaculture production was 0.82 million mt in 1992, half being derived from non-traditional species.

Marine Culture

Shrimp farming began in China in 1978 and special attention was paid to the culture of *Penaeus chinensis, P. japonicus, P. monodon and P. penicillatus.* The main form of culture utilized interconnected ponds occupying large areas. Shrimp farming developed so quickly that from 1978 to 1992, the area under culture increased 110 fold, eventually reaching 140 000 ha, and the output increased 460 fold, from 2 549 mt in 1980 to 200 000 mt in 1988, an annual rate of increase of more than 75%. The production value was 4 billion Chinese Yuan (US$ 800 million). Seventy percent of all the prawn farming operations were located in the Bay of Bohai Sea and two peninsulas. However, in 1993, a serious epidemic broke out; production decreased rapidly to 80 000 mt, and still continues to decline.

Because of their high economic potential, some marine fish (grouper, seabream, snapper, seabass etc.) also attracted the attention of fish farmers. These fish are cultured in cages or in indoor tanks. Because of the short history of this type of culture, output was not very high (less than 30 000 mt in 1987).

Shellfish culture in shallow areas has also developed very quickly. In 1950, the area under culture was 13 000 ha, with an output of less than 10 000 mt. By 1990, the area used for this type of culture had expanded to 350 000 ha, with more than 65% of available area being under production. The output of shellfish reached 700 000 mt in 1987, with about 65% of total production from marine culture.

The Aquaculture Situation in 1992

In 1992, total aquatic production was on the order of 15 million mt, of which 7.76 million mt was derived from aquaculture (52% of the total), 5.34 million mt from freshwater culture and 2.42 million mt from marine culture. The freshwater area under culture was 3.8 million ha (1.5 million ha of ponds, 2.3 million ha of lakes and reservoirs) of which 20 000 ha was for enclosed nets, 200 000 ha for industrial fish farming and 293 000 ha for special species culture. The area used for marine culture was 400 000 ha, with 140 000 ha being used for shrimp farming.

PRESENT STATUS OF AQUACULTURE HEALTH AND MAJOR PROBLEMS[1]

The status of aquaculture health in China was good before the 1980s. At that time, the pond culture of freshwater carps under low stocking densities was the major form of aquaculture practiced, and most of the common diseases seen in small ponds were basically under control, a result of research on fish pathology and the use of traditional health management strategies and control measures which were well established.

However, during the past 10 years losses due to pathogens have increased in both in extent and severity, along with the use of increased stocking densities, expansion of culture area, and the introduction of many new species and culture systems. To date, about 200 different diseases have been discovered; about 20 of these are serious diseases which can produce epizootics, sometimes with disastrous consequences for Chinese aquaculture. They have become the major factor hindering the development of aquaculture. For example:

Bacterial hemorrhagic septicemia of fish cultured in fresh water

This is an acute disease causing high mortality which is prevalent over more than half of China's mainland. It has threatened silver carp, bighead carp, golden carp, Chinese bream and common carp since the late 1980s. A loss of more than 100 million Chinese Yuan (US$ 20 million) was suffered annually for three consecutive

---

[1]Recent publications on diseases affecting Chinese aquaculture are given in Appendix I.

years. This disease subsided somewhat in 1993, but losses remained high at more than 60 million Chinese Yuan (US$ 12 million). The pathogens isolated were opportunistic bacteria common in many areas of the world: *Aeromonas punctata, Vibrio fluvialis and Yersinia ruckeri*; however, their virulence increased more than 10 or even 100 times in adverse circumstances.

Diseases of penaeid shrimp in ponds

Serious epidemic diseases appeared in south China in 1990 and spread to the whole of China in 1993, endangering *Penaeus chinensis, P. japonicus, P. monodon* and other species. More than 76% of shrimp ponds suffered reduced production, with a loss of some 120 000 mt of shrimp valued at up to US $ 400 million.

Hemorrhagic disease of grass carp

Hemorrhagic disease is the most serious viral disease of grass carp during the fingerling stage, with only 30% of the fish surviving to reach the market. An old saying *"become a millionaire if there is no epidemic in cultured fish"* points to this disease. To reduce the serious loss of fingerlings, an inactivated vaccine was prepared and mortality was lowered from 70% to 30% after injection. Several billion grass carp fingerlings are vaccinated every year, a method that is very time consuming. Some fish farmers do not pay attention to immunoprophylaxis, but hope to have a good medicine to cure the viral disease. Others ignore environmental hygiene and cause man-made epidemics. As a result, hemorrhagic disease still occurs in many regions and causes serious losses.

The current aquaculture health situation

- The number of common pond diseases is rising again, and new diseases of fish cultured in ponds (new pathogens, or pathogens spreading to new geographic areas or infecting new hosts) occur constantly.

There are about 40 common diseases which occur in fish ponds, of which ten, such as hemorrhagic disease, bacterial enteritis, bacterial gillrot, trichodiniasis, ichthyophthiriasis, sinergasiliasis etc. still prevail. As methods of prevention and cure were found, they were kept under control in the 1960s and 1970s. However, at present, many fish farmers do not take preventive measures by using medicine or vaccines, and mortalities have risen. For example, in Honghu Lake, the survival rate of grass carp fingerlings is only 30%. The prevalence of *Sinergasilus* reached 50% and morbidity due to enteritis and gillrot was above 50%. The situation in other regions was similar. This trend towards increased disease will continue if aquaculture health management is not emphasized. The appearance of epidemic diseases in large lakes and reservoirs makes their control very difficult.

Fish cultured in cages and enclosures are easily infected by pathogens causing epidemic diseases such as pox disease, myxosporidasis, dactylogyriasis, sinergasiliasis etc. These diseases are becoming much more common, and are causing serious losses. For example, an epizootic caused by *Ergasilus* broke out in a reservoir near Tianjing City, and several hundred common carp died in three or four days, producing a terrible smell which could be detected at 5 km distance. However, control of diseases in lakes and reservoirs is very difficult, as the areas involved are so large. Furthermore, there are no practical methods to control some of the diseases (*e.g.*, myxosporidasis). Sometimes waste water from factories is discharged into water bodies and causes mass mortalities.

- Many diseases appear in newly cultured species.

Newly cultured species are frequently affected by pathogens. The scale of the culture of these animals has developed rapidly, and most are kept at high stocking density and fed artificial feeds. There is no previous experience nor existing culture formats for these new species to assist farmers in keeping them healthy. This makes disease occurrence likely. We cannot even determine the cause of some diseases and do not know how to control them, thus aggravating the situation and hindering the development of aquaculture. Some important diseases are described below.

- In the 1980s, infectious diseases seriously affected pearl culture. The culture of *Hyriopsis cumignii*, a species which produces high quality pearls, almost disappeared became of its low resistance to disease. Farmers had to culture another species (*Cristaria plicata*) in its place, even though the quality and quantity of pearls it produced was not as good.
- Soft-shelled turtle is normally very resistant to disease, but even they suffer high mortality under the artificial conditions of culture.
- More than 60% of pond-cultured Chinese perch in Zhujiang Delta became ill and no effective control measures could be devised because these fish eat only live fish.
- The death of freshwater crabs in only one village in Hubei Province resulted in losses of about 6 million Chinese Yuan (US$ 120 000).

- Many epidemics occur in industrial fish farming.

Economic losses are usually serious and control difficult whenever diseases occur in industrial fish farming because fish are cultured in a high density environment. Furthermore, the poor quality of artificial feeds often leads to illnesses from nutritional or unknown causes. An example is obesity and exophthalmus of tilapia, which led to mortalities where no pathogen was found.

In summary, the situation with regard to aquaculture health has steadily deteriorated over the last 10 years, resulting in a variety of diseases which have endangered new areas and caused increased losses. These diseases are major limiting factors to aquaculture development, and it has been estimated that production would be 10-25% higher if they were controlled. The occurrence of disease is closely related to both the state of the local environment and culture patterns being used.

Major causes of health problems

- Deteriorating aquaculture environment and poor ecological conditions, leading to poor water quality (low pH and high NH_3 and NO_2 concentrations) and rapidly rising total amount of bacteria in water. Results are that on the one hand there is an increased susceptibility of cultured fish to disease, and on the other, the growth of pathogens is favored.

The causes are:

- Many new fish farmers lack experience in health management.
- Sludge from fish ponds in many regions is not cleared away in time for improvement of cultivation. In order to save money and manpower, some farmers even ignore basic preventive measures, such as pond disinfection.
- To achieve high profits, farmers culture fish at irrationally high densities. This overstocking leads to difficulties in maintaining good water quality.
- Over-exploitation of lakes and reservoirs, exceeding the ecological capacity.
- Industrial water pollution; some ponds near the suburbs are contaminated with chemicals.

- Lack of a good system of prevention, monitoring and quarantine.

As aquaculture developed, cultured species were frequently transferred from one region to another, carrying their pathogens with them. A serious shrimp disease which occurred in 1993 was obviously caused by the transfer of large numbers of *Penaeus japonicus* infected with viruses from southern to northern China. In many regions of northern China, these *P. japonicus* became diseased and died, with the disease spreading to *P. chinensis*. Those shrimp farms that were not in a position to exchange water from outside or which were unable to get fresh feeds from the epidemic area were only slightly influenced.

Another example is that in 1986, eyed eggs of rainbow trout were imported from Japan without strict quarantine. This importation lead to an outbreak of acute infectious disease (IPNV) in Shanxi Province causing 90-95% mortality.

- Irrational programs of development and culture patterns.

In some areas, several thousand ha of shrimp ponds or fish ponds are linked together. Some of them usually have no preventive installations for the treatment and disinfection of source and discharge water, so disease spreads widely once it appears in one or two ponds.

Protection of the environment was ignored when rapid large-scale development was planned. Waste water is drained from the ponds without treatment, causing pollution, which in turn, adversely affects culture conditions.

- Research conducted is inadequate to meet the demands of aquaculture development.

To some extent, inadequate basic research affects the subsequent application to aquaculture practice. Important areas where more research is needed include the study of the mechanisms of pathogenesis, the origin of pathogens, the distribution and epidemic patterns of diseases (especially viral diseases), the role of environmental stress in determining host resistance and pathogenicity, the prediction and monitoring of epidemics etc. Long-term research is needed. The use of some drugs is already limited or prohibited (*e.g.*, $HgNO_3$) and some effective methods for disease control in small ponds, such as hanging baskets or bags with drugs in pond water to prevent infection, are not suitable for use in big lakes and flowing water. Therefore research to develop new drugs and new methods of treatment is needed.

HISTORY AND PRACTICE: EXAMPLES OF AQUACULTURE HEALTH MANAGEMENT

Although Chinese freshwater aquaculture has a history of several thousand years, the scientific study of fish disease was begun only in the past few decades. However, people have accumulated some experience on health management. For example, in the polyculture of the four commonly raised carps, it is known that water quality is associated with parasitic infections causing diseases. In the 1950s and 60s, research on fish diseases expanded very quickly and a set of traditional health management systems was formed.

Recommendations include: the use of quick lime to disinfect ponds, the use of mixed compost (manure) to feed fingerlings and the hanging of baskets filled with bleaching powder or of bags filled with $CuSO_4$ at the feeding sites. All these methods are very easy to operate and are effective against fish diseases. Hanging drug bags or baskets are an ideal and practical method, as fish come into contact with the drug for disinfection. Over-dose is avoided, as the fish can keep away voluntarily if the concentration of drugs at the site is too high.

For prevention and cure of diseases, fish farmers follow the principle "prevention prior to illness and early treatment immediately after." They also carry out other efficient steps, the "three disinfections" and the "four rules." The implementation of these has proved indispensable to fish culture. For example:

- On regular use of drugs for prevention of diseases: Use drugs prior to the epidemic season of some fish diseases to reduce morbidity. For example, hanging bags filled with $CuSO_4$ and $FeSO_4$ against protozoans and filled with bleaching powder to prevent bacterial diseases.
- Clean the pond environment, including rapid discarding and burying of any dead fish in time to limit the spread of disease.
- Disinfect the tools and use them separately for individual ponds.

Three disinfection methods:

To clear and disinfect the ponds (either dry ponds or ponds with water) thoroughly using quick lime or bleaching powder.

To disinfect juveniles and fingerlings, use 3-4% salt, 10 ppm bleaching powder, 8 ppm $CuSO_4$ or 20 ppm $KMnO_4$ for 20 min.

To disinfect feeds and feeding sites (feeds should be clean, and kept fresh and alive), immerse plant feeds in 0.05% bleaching powder for 20-30 min; disinfect 300 kg manure fertilizer with 120 g of bleaching powder or apply into the water after complete fermentation; disinfect the feeding sites by hanging baskets filled with bleaching powder and by casting the disinfectant around them.

Four rules in feeding:

Choose the quality (either dry or fresh) and the amount of feeds; fix the feeding site and time.

Efficient control methods have been found for many fish diseases that often caused serious losses in the past. Diseases such as gillrot, enteritis, hemorrhagic disease, ichthyophthiriasis, fungus disease etc. in ponds are now under control. The history of research development in fish diseases is almost the same in China as in other countries. At first, research was carried out on parasites, then the focus turned to bacterial and viral diseases.

In brief, prevention is the major element in traditional aquaculture health management. It ensures the minimum possibility for fish to come into contact with pathogens and become infected.

STATUS OF INNOVATIVE AQUACULTURE HEALTH MANAGEMENT IN CHINA AND COMPARISON OF ITS EFFECTIVENESS WITH TRADITIONAL MANAGEMENT

Generally speaking, the policy of "putting prevention first" is still maintained in present health management. It conforms with the changing aquaculture situation, including the extended scale of culture, the increased production and the rise in stocking densities, and also, the occurrence of many new viral and bacterial diseases. Of course, the advantages and disadvantages of these measures and policies still need further evaluation through practice.

Improving the pattern of culture in ponds

The intensification of aquaculture practices which began in the late 1970s has had a great impact on the environment, causing disease epidemics and affecting the disease resistance of fish. New culture methods stressing high production from the ponds were adopted, modifying the existing traditional culture patterns.

In the past, fish farmers never fed pond-reared fish during the winter and kept the fish under high density. Studies indicated that fish still require adequate nutrition and energy in winter. In the new culture method, fish are kept in the original ponds at low density, reducing injuries to the body surface, and feeds are replenished on sunny days, resulting in healthier fish and enhancing their resistance to diseases during winter.

In the traditional polyculture method, the ratio of grass carp to silver carp (and/or bighead carp) was 1:3. However, this ratio was considered out of proportion, as the efficiency of energy transformation was not high enough. To meet the demands of silver and bighead carp, pond water had to be fertilized, often resulting in bad water quality and inducing disease in grass carp. Now the ratio has been changed from 1:3 to 1.5-2.5:1, a measure which can efficiently raise survival rates, increase production and improve water quality. Using these new methods, production rose from 690 kg/ha in 1978 to 2070 kg/ha in 1988.

Immunization against viral and bacterial diseases

In the past, the prevention of fish disease was mainly directed towards eliminating infectious sources and severing the routes of transmission. Now, more attention is paid to enhancing the general level of health and increasing disease resistance. With parasitic diseases almost under control, viral and bacterial infections are becoming increasing prominent. There are no drugs to cure viral infections. Immunological methods of prevention, such as the use of inactivated vaccine against hemorrhagic disease, and bacterial vaccines against gill-rot, enteritis, hemorrhagic septicemia, vibriosis etc., have been developed.

The use of vaccines in aquaculture has been approved and accepted by fish farmers and the results are fruitful. The methods used in the preparation of vaccines have also advanced, beginning with inactivated vaccines, and then to live vaccines, sub-unit vaccines and gene vaccines, the aim being to increase the specificity, effectiveness and safety. However, it was found that after fish were immunized with inactivated vaccine against hemorrhagic disease, the morbidity of some bacterial diseases, such as enteritis and gillrot, was also clearly reduced. It is suggested that the impurity of the vaccine plays a beneficial role by enhancing the nonspecific immune response of fish. Perhaps vaccines prepared from the organs of sick fish were actually a mixture against viruses and bacteria. Inactivated vaccines prepared from organs of sick fish have been well accepted by fish farmers. Effectiveness, convenience and low cost are the most significant requirements.

Another trend is the use of additive drugs such as peptidoglycan (PG) to fortify the immunity of fish.

Use of new drugs in aquaculture

Many new drugs have been produced for disease control in aquaculture. These include various disinfectants *e.g.*, PVP-iodine, CDB Cleraron (sodium dichloroisocyanurate) etc.; and various new types of antibiotics for oral use.

New drugs are undoubtedly more effective in killing pathogens than are those previously available. However, they themselves possess certain advantages and shortcomings and, in practice, their use must be individually evaluated. For example, CDB Cleraron is highly effective, convenient to transport and can clean ponds rapidly, fish being able to be stocked into disinfected ponds only a few days after treatment. If lime is used, 10 days are required. However, the former lowers the pH of water and cannot improve the quality of the water or the pond bottom sludge, so for long-term use, its effectiveness has yet to be fully evaluated.

The excessive use of antibiotics often leads to reduced resistance (*e.g.*, fungal diseases may occur easily) and to formation of drug-resistant bacterial strains or the killing of beneficial bacteria.

Establishment and application of rapid detection methods

Because bacteria and viruses cannot be detected visually, new detection techniques developed through immunology and molecular biology are used, as these new methods are more reliable, sensitive and rapid. They may be the basis for enforcing quarantine rules. A number of research institutes are developing modern methods to detect pathogens of cultured species, such as IPN virus in rainbow trout and IHHNV in shrimp. However, these advanced methods are used only in research institutes and some quarantine departments and have not yet been widely adopted in all regions of China. Visual inspection is still widely used for on-site diagnoses, relying on the experience of the diagnostician.

Prevention of diseases by ecological methods

Researchers have paid attention to the relationship between the environment, fish health and disease. They realize that ecological control is the best way to combat potential pathogens and have begun to exploit new culture formats (*e.g.*, relationships between density, benefit and health; the polyculture of shrimp and fish). A good example is the successful culture in water with low salinity of *P. monodon* carrying viral infections in their tissues.

Study of disease-resistant strains and specific pathogen-free fish

These studies are still underway, and it will be some time before their application is realized in China.

RESEARCH INSTITUTIONS INVOLVED IN PROMOTING AQUACULTURE HEALTH MANAGEMENT

The Bureau of Fishery, Ministry of Agriculture and subordinate Bureaus of Fishery in provinces and cities are in charge of fishery management, including disease control at the national level. They form an administrative system of management, and are responsible for planning fishery development and the scale of culture, circulating information of epidemic diseases in the various regions, organizing experts to formulate a policy for quarantine, arranging for institutes to tackle key problems etc.

Scientific research is being carried out by the following agencies and institutes:

Institutes of the Chinese Academy of Science (CAS): Institute of Hydrobiology, Institute of Oceanology, South China Sea Institute of Oceanology etc.

The Chinese Academy of Fishery Science (CAFS), which was founded in 1982, has a number of subordinate institutes: Yangtze Institute, Zhejiang Institute, Heilongjiang Institute, Center of Freshwater Fishery, Yellow Sea Institute, East China Sea Institute, South China Sea Institute, Zhujiang Institute etc.

Departments and research units of colleges and universities (there are groups or laboratories working on fish diseases, especially at many fishery colleges).

The Fish Disease Section of the Chinese Fishery Society (FDS/CFS) was set up in December 1985. Its main tasks are to carry out academic discussion and exchange, to organize symposia on special subjects, to offer technological training on the control of fish diseases, to edit and publish academic journals and popular science articles, and to initiate communication activities both at home and abroad. The Fish Disease Section holds national academic symposia every four years.

A network for fish disease control was founded in December 1990. It not only transmits and forecasts information concerning fish diseases, but also communicates information on new technologies, new drugs and new achievements in the field of disease control; offers various training courses; provides technology and consultation services; and coordinates the relationship between scientific research, production and concerned enterprises.

The major sources of current funds for fish health research:

Some key projects form part of a five-year plan for China, for example, studies on hemorrhagic disease in grass carp and epizootics in freshwater cultured fish. The subjects chosen must make important contributions to the development of the national economy, and have high applicability and feasibility.

The National Science Foundation of China (NSFC) and the Science Foundations of some provinces support basic research and new projects in science.

FUTURE PROSPECTS AND RECOMMENDATIONS

Aquaculture development should be based on disease prevention and on maintaining ecological balance

Health problems caused by the irrational distribution and over development of aquaculture cannot be solved by creating new medicines, but only by transforming aquaculture itself. The pattern and technology of culture should be improved so as to reduce stress on the environment and decrease risks to production (*e.g.*, changing from a large area of single species production to multi-species production). In planning future development, we must readjust the amount and structure of culture, defining the culture scale according to levels sustainable by the environment in the area of the aquaculture development.

Research and development on diagnostic technology

We need two different types of detection techniques, which together, can form a preventive system covering multiple levels. One group of techniques is needed for on-site diagnosis to help fish farmers quickly determine the causative agents of problems, while the other groups are those techniques which should be applied in laboratories, and have high sensitivity and specificity for use in quarantine and the prediction of epidemic diseases.

Formulation of quarantine laws, improvement of detection to prevent pathogens from spreading artificially, and licensing of production

Our research should be directed towards the study of those diseases which have the highest economic impacts. Researchers should be strongly encouraged to study epidemic diseases occurring among new culture species, in cage culture, and in high density culture in big lakes and reservoirs and the marine environment.

Enhancement of fish farmers' knowledge

For example, the preventive measures taken against diseases in ponds in China are effective. The key problem is how to have them executed and popularized. Many fish farmers do not realize the relationship between the culture environment and disease. They accept that disease results only from the existence of pathogens and place their hopes in new medicines. Therefore, it is most important to raise their level of knowledge by offering training courses on fish health and preventive methods.

Enhancement of international cooperation

International cooperation should be increased in such areas as the exchange of standard antisera and antigens, the standardization of diagnostic procedures and the monitoring of diseases.

APPENDIX I

Some Recent Publications on Diseases in Chinese Aquaculture Systems

Chen Chileu and Pan Jinpei, 1992. Chapter 13, Prevention and control of fish diseases. *In* Cultivation of Chinese Freshwater Fishes (3rd edn.). Liu Jiankang and He Biwu (eds.) Science Press, Beijing, p. 519-629.

Chen Yanxin, Chen Hong, Han Xianpu and Li Wei, 1992. A viral epidemic disease of eel occurred in China. Chinese Sci. Bull. 23: 2910-2912.

Han Xiangpu and Li Wei, 1989. Study on edwardsielliasis of the eel. Acta Hydrobiol. Sinica, 13: 259-264.

Han Xianpu and Li Wei, 1993. Studies on the hemorrhagic open-mouth disease of *Anguilla japonica*. Acta Hydrobiol. Sinica, 17: 285-287.

Jiang Yulin and W. Ahne, 1989. Some properties of the etiological agent of the hemorrhagic disease of grass carp and black carp. *In* Viruses of Lower Vertebrates. W. Ahne and E. Kurstak (eds.) Springer-Verlag Publ., Berlin.

Jiang Yulin, Li Yan and Li Zhengqiu, 1991. Electron microscopic observation of pathogen of carp-pox disease. Acta Hydrobiol. Sinica, 15: 193-195.

Jiang Yulin, Li Zhengqiu, Han Guangshu, Wang Xue and Gong Xianghong, 1995. Electron microscopic observation of bacilliform virus in shrimp (*Penaeus orientalis*). Trans. Res. Fish Dis. No. 2. China Ocean Press, Beijing.

Jiang Yulin, Yu Ping and Li Zhengqiu, 1990. Rapid detection of infectious pancreatic necrosis virus (IPNV) with the enzyme-linked immunosorbent assay (ELISA). Acta Hydrobiol. Sinica, 14: 276-279.

Meng Qingxian, 1991. Handbook of the Prophylactics and Therapeutics of Shrimp Diseases. Ocean University of Qingdao Publication, Qingdao.

Pan Jinpei (ed.), 1988. Handbook of the Diagnosis, Prophylaxis and Therapeutics of Fish Diseases. Shanghai Scientific and Technical Publishers, Shanghai.

Xu Bohai, Yin Zhan, Wu Yushen and Cai Taozhen, 1991. Studies on an outbreak of a new epizootic in silver carp and bighead carp - *Yersinia ruckeri*, a new pathogen of silver carp and bighead carp. Chin. Sci. Bull. 36: 1825-1828.

Xu Bohai, Yin Zhan, Wu Yushen and Cai Taozhen, 1993. Studies on the taxonomy of pathogenic bacteria of the bacterial hemorrhagic septicemia in cultured fishes in fresh water. Acta Hydrobiol. Sinica, 17: 259-266.

Ye Xueping, Yang Guangzhi, Luo Yizhi, Chen Yinliang and Chen Zhihong, 1992. Large-scale culture of grass carp cell and virus by using bioreactor. J. Fish. China, 16: 1-6.

Yin Zhan and Xu Bohai, 1994. Observation of *Vibrio fluvialis* Biovar III in tissues of infected silver carp (*Hypophthamichthys molitrix*) using the fluorescent antibody technique. Acta Hydrobiol. Sinica, 18: 95-96.

An Overview of Health Management of Coldwater Fish and Shrimp in Japanese Aquaculture

K. Inouye

National Research Institute of Aquaculture
Tamaki, Mie 519 04, Japan

Inouye, K. 1996. An overview of health management of coldwater fish and shrimp in Japanese aquaculture. *In* Health Management in Asian Aquaculture. Proceedings of the Regional Expert Consultation on Aquaculture Health Management in Asia and the Pacific. R.P. Subasinghe, J.R. Arthur & M. Shariff (eds.), p. 104-114. FAO Fisheries Technical Paper No. 360, Rome, FAO. 142 p.

Abstract

Disease problems has been an important consideration in Japanese aquaculture over the past decade. Most of the disease outbreaks are caused by the environmental problems stemming from over stocking and entry of new exotic pathogens. In 1992 alone, Japan lost US$ 264 million due to diseases in aquaculture. Japan's approach to aquaculture health management consists of three major components: provision of facilities for diagnosis and prevention, implementation of appropriate rules and regulations, and improvement of infrastructure. This paper describes the manner in which health management of coldwater cultured fish and shrimp is being conducted.

INTRODUCTION

In Japan, aquaculture production in 1992 was 1 397 000 metric tonnes (mt), worth US$ 8.4 billion, and amounting to 15.7% of the total fishery production. Aquaculture production of coldwater finfish and shrimp in 1992 was 47 091 mt, or 3.4% of the total aquaculture production. The major species of cultured coldwater finfish and shrimp are coho salmon (*Oncorhynchus kisutch*), which are cultured in fresh water until smoltification and then raised in sea water, kuruma shrimp (*Penaeus japonicus*), which are reared in marine systems, and rainbow trout (*O. mykiss*) and Japanese native salmonids (landlocked salmon, *O. masou*; dwarf rill trout, *O. rhodurus*; and mountain trout, *Salvelinus pluvius*), which are raised in fresh water. Coho salmon is a species whose production has increased rapidly in the last 10 years. In 1992, it was the highest in production, with 25 519 mt, followed by the freshwater salmonids and kuruma shrimp, with 19 385 mt and 2 187 mt, respectively.

The very rapid development of aquaculture has been accompanied by disease problems. These problems are partly caused by environmental deterioration and overstocking, which lead to health disorders in cultured fish, and are partly due to the entry of exotic pathogens.

Table 1 shows the major diseases of fish and shrimp, their causative agents, and the year of their first occurrence. Infectious pancreatic necrosis (IPN), infectious hematopoietic necrosis (IHN), erythrocyte inclusion body syndrome (EIBS) and bacterial kidney disease (BKD) have been introduced from abroad with imported eggs. Recently, mass mortalities caused by a new baculovirus occurred in kuruma shrimp farms in Japan. The causative virus of this disease was also introduced from abroad.

The Fisheries Agency regularly evaluates the damage caused by fish disease in intensive aquaculture. Each year the Agency has prefectural governments send out questionnaires to all aquaculturists and inform the Agency of the results. The damage caused by disease in finfish and shrimp aquaculture in 1992 was estimated to be 18 269 mt, worth US$ 264 million, compared to a total production of 356 154 mt, worth US$ 4 376 million. Thus, about 5% of the total production of finfish and shrimp was lost to disease in 1992.

Losses due to disease suffered by coldwater finfish and shrimp culture (% total production) were on the order of 2.5% for coho salmon, 6.2% for rainbow trout, 5.3% for Japanese native salmonids and 13.3% for kuruma shrimp. The loss of coho salmon due to disease in 1992 was estimated to be 733 mt, worth US$ 4.6 million. EIBS caused the largest loss (289.1 mt), followed by vibriosis (197.7 mt) and BKD (162.9 mt) (Table 2). The loss of rainbow trout in 1992 was estimated to be 630.2 mt, worth US$ 6.2 million. IHN caused the largest loss (199.3 mt), followed by saprolegniasis (109.0 mt), bacterial gill disease (BGD) (64.7 mt), ichthyophonosis (51.7 mt) and vibriosis (42.9 mt) (Table 3). Losses of Japanese native salmonids were estimated to be 266.4 mt, worth US$ 1.8 million. Furunculosis caused the largest loss (95.6 mt), followed by saprolegniasis (27.5 mt), vibriosis (26.0 mt), ichthyophthiriasis (white spot disease) (22.0 mt) and BGD (18.9 mt) (Table 4).

Vibriosis caused the largest loss in kuruma shrimp production in 1992, the damage amounting to 149.5 mt, worth US$ 26.8 million. In 1993, a new baculovirus disease occurred in major kuruma shrimp farms in Japan. The loss of shrimp caused by this disease was estimated to be 320 mt, worth US$ 24 million.

DIAGNOSIS AND PREVENTION

Diagnosis of fish diseases is made mainly by staff of fisheries experimental stations or of the fish diseases control centers of the prefectural governments. A small number of staff at the Fisheries Co-operation Association who have followed a training course in fish disease control also make diagnoses.

When a disease breaks out, chemotherapy is important. At present, 29 antibacterials are permitted for fisheries usage in Japan: 5 sulfa drugs, 2 nitrofurans, 7 other synthetic antimicrobial drugs, and 15 antibiotics. Oxytetracycline hydrochloride (OTC), oxolinic acid (OA), sulfamonomethoxine (SMM) or its sodium salt (SMM-Na), and florfenicol (FF) are used to treat bacterial diseases of coho salmon. OTC, OA, sulfadimetoxine (SDM) or its sodium salt (SDM-Na), SMM, SMM-Na, FF and sulfisozole (SIZ) are used for rainbow trout; OTC, OA, SMM, SMM-Na, FF and piromidic acid (PA) for dwarf rill trout; and OTC and OA are used for treating kuruma shrimp (Table 5).

Only one commercial vaccine for salmonid vibriosis has been approved for use in Japan. The bath vaccination method is currently employed.

Several germicides are used in aquaculture for the prevention of diseases such as IHN, IPN and vibriosis (Table 6), iodophor being the most commonly used. Living salmonid eggs are disinfected by iodophor immediately after importation. When eggs are introduced from a domestic hatchery, the fish culturist uses germicides for their disinfection in the same way. Farmers use several kinds of invert soaps, cresol, sodium hypochlorite and bleaching powders to sanitize aquacultural materials and facilities. Recently, some new methods, such as treatment using ultraviolet light and ozone, are employed in some hatcheries for sanitizing the water supply.

RULES AND REGULATIONS

Quarantine

No regulations for fish quarantine have been established in Japan. However, to avoid the entry of pathogens from abroad, the Fisheries Agency has requested that certain steps be taken through the prefectural governments. Namely, at importation of live eggs or juveniles, the importers are requested to a) attach a health certificate issued by the governmental authority of the exporting country stating that the eggs or juveniles are not infected, b) have juveniles or eggs examined for pathogens at the time of importation, and c) disinfect juveniles or eggs immediately after importation. Examination for pathogens at the time of importation is entrusted to the Japanese Fisheries Resource Conservation Association (JFRCA) by the Fisheries Agency. Voluntary inspections of eggs and fingerlings are performed for viruses, bacteria and parasites by JFRCA in cooperation with universities and the National Fisheries Research Institute.

Information on disease occurrence

To prevent the spread of pathogens within the country, measures are taken by Fishery Experimental Stations and Fish Disease Control Centers. On request, they examine for pathogens and make diagnoses when juveniles and eggs are transferred between prefectures. Prefectural governments gather information on the occurrence of diseases from aquaculturists and Fisheries Co-operative Associations and distribute this information to other Fisheries Co-operative Associations and prefectural governments as the occasion demands. A computerized network linking the Fisheries Agency, JFRCA and the prefectural governments is being established.

Restrictions on the use of drugs in aquaculture

All drugs should be controlled by the Central Pharmaceutical Affairs Council before being approved for manufacture by the Ministry for Agriculture, Forestry and Fisheries. The examination is based on the *Pharmaceutical Affairs Law* for the safety and efficiency of medicines. Usage of fishery medicines is also regulated by *The Standard to be Observed by User*, which is based on the *Pharmaceutical Affairs Law* and indicates the administration method, dosage, withdrawal period etc.

INFRASTRUCTURE

Institutions

- ***The Fisheries Agency***

The Fisheries Agency belongs to the Ministry of Agriculture, Forestry and Fisheries, and has jurisdiction over fisheries administration, including disease control and aquaculture promotion. The Fisheries Agency has nine national research institutes and the Shimonoseki University of Fisheries. These facilities carry out research or training in fisheries.

The National Research Institute of Aquaculture (NRIA) belongs to the Fisheries Agency, and was established in 1979 to promote basic research on aquaculture. There are five research divisions in NRIA, one of which is the Fish Pathology Division. This division has 12 researchers working in four research sections: the Pathogen Section, the Pathophysiology Section, the Pharmacology Section and the Immunology Section.

Since 1973, when losses due to disease began to be recognized as a serious problem to aquaculture, the Fisheries Agency has practiced systematic measures for the prevention of fish disease. These measures can be roughly divided into two parts: one is to support the work carried out by the prefectural governments and the Fisheries Co-operative Associations, and the other is the work entrusted to the Japan Fisheries Resource Conservation Association. The Fisheries Agency is taking broad countermeasures for the prevention of fish diseases by working together with the prefectural governments, fishermen's cooperatives and aquaculturists.

- ***Prefectural Governments***

The fishery sections of prefectural governments exercise jurisdiction over fisheries administration mainly within their prefectures and coastal areas. As for disease prevention, measures taken at the actual aquacultural sites are very important. Prefectural governments play a very important role in guiding farmers, either directly or through the fishermen's cooperatives.

There are a total of 96 Fishery Experimental Stations in Japan, each prefecture having at least one station. At each station, some divisions of carry out research for aquaculture and the control of fish diseases. In most prefectures where disease causes severe damage to aquaculture, there are Fish Disease Control Centers (20 in total), either as a part of the Fishery Experimental Stations or as independent organizations. Prefectural governments have also established Fisheries Extension Offices to give technical guidance to fishermen. These offices and the Fishery Experimental Stations work together to take preventive measures against fish diseases.

The total number of staff of prefectural governments engaged in disease prevention was 470 in 1992. In addition, 30 staff from cities, towns and villages are also engaged in these measures.

- ***Fisheries Co-operative Associations***

Fish farmers often establish their facilities quite close to each other. In marine aquaculture, farms are usually set up within a rather confined area, such as a bay, and in freshwater aquaculture, farms often utilize the same source of water. Therefore, the Fisheries Co-operative Association, which gives guidance to fishermen and controls the local fishing ground, plays an important role in fish disease prevention. However, many of these associations are on a very small scale and thus the level of guidance varies greatly according to the association. The total staff of all associations engaged in disease prevention measures in 1992 was only 80.

- ***Japan Fisheries Resource Conservation Association (JFRCA)***

Since 1984, the Fisheries Agency has carried out some activities by entrusting JFRCA to establish uniform techniques for disease prevention. The undertakings involve promotion of measures for disease prevention, technical extension, training, improvement of diagnostic methods and voluntary inspection of imported living fish and eggs for pathogens. In carrying out these undertakings, JFRCA also receives advice and support from the National Research Institute of Aquaculture, the universities and the fishery experimental stations.

- *Universities*

Seventeen universities having fishery faculties (including the Shimonoseki University of Fisheries belonging to the Fisheries Agency), and another 16 universities with veterinary faculties have educational research facilities for fish disease.

Establishment

The Fisheries Agency provides financial support to the prefectural governments and the Fisheries Co-operative Associations for establishing facilities for disease prevention.

Training

At the farm level, fish disease specialists are indispensable in order to carry out disease prevention measures, such as providing advice to farmers, making diagnoses, and disinfecting facilities and equipment. Therefore, since 1973 the Fisheries Agency has trained staff from prefectural governments and, since 1984, staff from the Fisheries Co-operative Association have been included in this training. This training is aimed at educating fish disease specialists and improving their quality. In recent years, the Fisheries Agency has entrusted this training to JFRCA.

In Japan, the training of fish disease technical workers is carried out separately in both fishery and veterinary fields, with aquacultural guidance and leadership being mainly carried out in the fishery field. There are several training courses. A basic course consists of lectures and practical exercises given 20 days each year for a duration of three years. In 1989, there were 30 to 40 trainees for each yearly course; altogether 100 staff participated in the training. Other training for fish disease technical workers is also carried out, depending on the circumstances. Since 1974, the Japan Veterinary Medical Association has individually carried out annual fish disease training via post-graduate education of veterinarians.

Extension

Countermeasures for disease prevention need to be uniformly applied. To achieve this, extension activities with respect to disease countermeasures are implemented as follows: JFRCA distributes textbooks on diagnosis and techniques of disease prevention and lends out movies and VTR to prefectural staff who are in charge of guiding farmers. The prefectural governments also offer courses and distribute textbooks and pamphlets on fish diseases. These steps are taken to extend knowledge on fish disease, disease prevention and the use of chemotherapeutants.

National programs and development plans

Disease prevention measures should be taken cooperatively by the national and prefectural governments, the Fisheries Co-operative Associations and aquaculturists. In addition, disease control systems for the farms and a rapid and appropriate response by farmers to disease problems are important.

To promote an efficient disease prevention plan, the Fisheries Agency, prefectural governments and Fisheries Co-operative Associations hold meetings on disease prevention. These meetings cover the regulations and administrative policy of each agency or organization and are designed to ensure that the prefectural organizations and farmers follow the overall administrative plans of the Fisheries Agency.

The following activities should be undertaken by the Fisheries Agency to promote fish disease prevention:

- Studies are needed to elucidate the mechanisms and the causes of fish diseases, and to develop disease prevention techniques such as vaccines.
- New medicines applicable to different species of fish, and to diseases that are resistant to medicines currently used in aquaculture need to be developed.

Husbandry techniques for fish health management by farmers need to be improved, and a fish-disease control system by the Fisheries Co-operative Associations should be established.

Table 1. Major diseases of coldwater fish and shrimp in Japanese aquaculture.

Disease	Year[1]	Causative agent	Host species
streptococcicosis	1977	*Streptococcus* sp. (¦À-hemolytic)	salmonids
vibriosis	1964	*Vibrio anguillarum* (*Vibrio* sp. PJ)	salmonids (shrimp)
BKD	1973	*Renibacterium salmoninarum*	salmonids
furunculosis	1929	*Aeromonas salmonicida*	salmonids
IHN	1971	IHN virus	salmonids
IPN	1964	IPN virus	salmonids
EIBS	1986	EIBS virus	salmonids
"RV-PJ" infection	1993	RV-PJ (Baculovirus)	shrimp

[1]Year of first occurrence.

Table 2. Losses of coho salmon due to disease in 1992.

Disease	Losses in Volume (mt)	Losses in Value (US$ x 1000)
EIBS	289.1	1578.8
vibriosis	197.7	1370.6
BKD	162.9	1064.7
herpesvirus	11.8	67.1
furunculosis	4.5	54.1
saprolegniasis	0.1	1.2
others	67.0	438.8
Total	**733.1**	**4575.3**

Table 3. Losses of rainbow trout due to disease in 1992.

Disease	Losses in Volume (mt)	Losses in Value (US$ x 1000)
IHN	199.3	3331.8
saprolegniasis	109.0	708.2
BGD	64.7	598.8
ichthyophonosis	51.7	312.9
vibriosis	42.9	297.6
IPN	4.2	160.0
columnaris disease	11.7	117.6
furunculosis	9.4	91.8
BKD	6.9	52.9
ichthyophthiriasis	1.1	28.2
streptococcicosis	2.5	22.4
others	126.8	456.5
Total	**630.2**	**6178.7**

Table 4. Losses of Japanese native salmonids due to disease in 1992.

Disease	Losses in Volume (mt)	Losses in Value (US$ x 1000)
furunculosis	95.6	568.2
EIBS	9.2	227.1
BGD	16.0	174.1
saprolegniasis	27.5	135.3
IHN	18.9	120.0
ichthyophthiriasis	22.0	57.6
coldwater disease	0.9	52.9
BKD	5.9	41.2
vibriosis	26.0	27.1
streptococcicosis	7.4	14.1
others	37.0	371.8
Total	**266.4**	**1789.4**

Table 5. Major drugs used to treat cultured coldwater fish and shrimp in Japan.

Species	Drug	Administration and dosage	Withdrawal period
Coho salmon	Oxytetracycline hydrochloride	Orally, mixed in food, < 50 mg (potency)/kg fish/day	30 days
	Oxolinic acid	Orally, 20 mg /kg fish/day	21 days
	Sulfamonomethoxine or its sodium salt	Orally, 100 mg /kg fish/day	30 days
	Florfenicol	Orally, 10 mg/kg fish/day	14 days
Rainbow trout	Oxytetracycline hydrochloride	Orally, 50 mg (potency) /kg fish/day	30 days
	Oxolinic acid	Orally, 20 mg /kg fish/day	21 days
	Sulfadimethoxine or its sodium salt	Orally, 100 mg /kg fish/day	30 days
	Sulfamonomethoxine or its sodium salt	Orally, 150 mg /kg fish/day	30 days
	Sulfamonomethoxine sodium salt	Bath, dissolving <10 kg/mt of salt water (less 1%)	15 days
	Florfenicol	Orally, 10 mg/kg fish/day	14 days
	Sulfisozole or its sodium salt	Orally, 200 mg/kg fish/day	15 days
Japanese native salmonids	Oxytetracycline hydrochloride	Orally, 50 mg (potency) /kg fish/day	30 days
	Oxolinic acid	Orally, 20 mg /kg fish/day	21 days
	Sulfamonomethoxine sodium salt	Bath, dissolving <10 kg/mt of salt water (less 1%)	15 days
	Florfenicol	Orally, 10 mg/kg fish/day	14 days
	Piromidic acid	Orally, 20 mg/kg fish/day	20 days
Kuruma shrimp	Oxytetracycline hydrochloride	Orally, 50 mg (potency) /kg shrimp/day	25 days
	Oxolinic acid	Orally, 50 mg /kg shrimp/day	30 days

Table. 6. Disinfectants used for prevention of fish diseases in Japanese aquaculture.

Germicide	Subject Disinfected
propanol	hands
ethanol	hands
cresol	hands, shoes
sodium hypochlorite	tanks, ponds, facilities
bleaching powder	tanks, ponds, facilities
iodophor	eyed eggs, hands, equipment
benzalkonium chloride	hands, shoes, cars, nets, equipment
benzethonium chloride	hands

Aquaculture Health Management in Singapore: Current Status and Future Directions

Frederic H.C. Chua

Fish Health Section, Animal And Plant Health Inspection Division
Primary Production Department, Singapore.

Chua, F.H.C. 1996. Aquaculture health management in Singapore: current status and future directions. *In* Health Management in Asian Aquaculture. Proceedings of the Regional Expert Consultation on Aquaculture Health Management in Asia and the Pacific. R.P. Subasinghe, J.R. Arthur & M. Shariff (eds.), p. 115-126. FAO Fisheries Technical Paper No. 360, Rome, FAO. 142 p.

Abstract

Although the total area utilized for aquaculture in Singapore is comparatively small, it annually produces over US$ 30 million worth of ornamental fish and marine foodfish. It has been predicted that, with the introduction of the Agrotechnology Parks concept, the value of aquaculture production in Singapore will reach US$ 42 million by the year 2000. This paper describes the approach taken by the Government of Singapore to ensure production of fish through hygienic methods employing better planned management systems.

INTRODUCTION

The ornamental fish and marine foodfish sectors make up the aquaculture industry in Singapore. Together they produce over US$ 30 million worth of farmed fish a year. The production from the ornamental fish sector accounts for about 75% of this total value. Freshwater foodfish production is insignificant.

Singapore exported a total of US$ 57 million worth of ornamental fish in 1994. This consisted of imported as well as locally farmed fish produced from 107 freshwater farms which cover a land area of 170 ha. By market demand, a diversity of species is cultured. More than 300 varieties can be found, comprising the live bearers (viz., guppy, platys, mollies, swordtail) and egg layers (angelfish, tetras etc.). Most farms culture a few species at a time. Due to competition, there has been a move to concentrate on and upgrade the farming and breeding of a few high-value species such as the varieties of discus and dragon fish (arowana). Nevertheless, the various guppies still form the bulk of production.

While many farms still employ what are considered to be traditional methods, such as growing fish in organically fertilized earthen ponds, there is a shift towards more hygienic production methods employing better planned management systems. This is seen in the establishment of farms within Agrotechnology Parks. In these sites, which have been set up by the Primary Production Department (PPD), land use is optimized through improved site planning and engineering. An increasing number of farms are being located at Agrotechnology Parks, and by the year 2000, a total production area of 260 ha from the Parks is expected to increase production to US$ 42 million a year. As the intensity of farming would increase with this development, the industry would be less reliant on imported fish for its export trade.

Production periods for ornamental fish are generally short, being between one to three months. Most stock are farm bred. Grading and sexing are routinely done during the production cycle.

Farmers use a variety of farm-prepared feeds. Common ingredients used include wheat, oats and fish meal. Live feed, such as *Moina* and tubifex worms, form part of the diets of some age groups (*e.g.*, fry, broodstock). Commercial pellets such as eel mash and tilapia pellets are also used.

Mariculture development took off after the Marine Fish Farming Scheme was implemented in March, 1981. There are now 79 floating marine netcage farms off the northern coast of Singapore (Johore Straits), which produce over 2000 metric tonnes (mt) of marine foodfish annually. The species farmed include seabass, groupers, siganids, snappers, threadfin, mud crabs, lobsters etc. Perhaps driven by good market demands for live fish in Hong Kong, Taiwan and China, and increasing production costs locally (particularly labor costs), many farms are switching to the transshipment trade. Here market-sized fish are imported from Indonesia, Malaysia and India, held for one to four weeks, and re-exported live. Because of the shorter turnover period, overall risk and production costs are considered to be lower.

The farms stock a variety of species and age groups at a time. Since both transshipment and growout are often carried out simultaneously, the culture periods vary widely, being between one to eight months. Although there are no particular periods of the year during which the introduction of fingerlings for on-growing is preferred, farmers generally avoid stocking fish during the monsoon seasons (June-July and November-January) due to higher post-stocking losses. Most feeding is with trash fish, which is supplied daily from the local port. However formulated diets are slowly finding their way into more farms.

The bulk of fingerlings stocked for growout comes from neighboring countries. However, there is an active program to commercialize the breeding and larviculture of species considered to have better market potential, such as the groupers. A project undertaken by the PPD in conjunction with the National University of Singapore (NUS) aims to produce grouper fry under captivity. There are currently three commercial hatcheries in operation, producing fry of seabass and snappers to supply the local and export markets. With government support, successful breeding of the fourfinger threadfin (*Eleutheronema tetradactylum*) has been accomplished in one of the hatcheries.

Constrained by limited available coastal areas for farming, mariculture will become more intensive in the future. Coupled with inherently high production costs relative to neighboring countries, increased attention will be given to the more highly priced species, such as groupers and threadfin. The focus is to enhance the development of commercial hatcheries, thereby making available a wider variety of fry for export. New sites for coastal aquaculture will also be developed, such as near the southern islands of Singapore, where water quality is expected to be better and where larger farming units with deeper netcages can be used.

HEALTH PRACTICES IN SINGAPOREAN AQUACULTURE

Some health practices commonly carried out by ornamental fish farms are listed in Table 1. Many ornamental fish farms also practice on-farm quarantine. Newly introduced fish are kept reasonably isolated for a few days to a week before stocking into the rest of the farm. Some farms, such as guppy farms which have developed their own varieties, usually do not find it necessary to introduce stock, as they already have an adequate breeding pool.

There appears to be a difference between farms in the Agrotechnology Parks and those which are not, in disease control measures used and in the rate of success of these measures. In contrast to the more traditional farms, those in Agrotechnology Parks have layouts that facilitate disease control. During a disease outbreak among guppies in 1993 which was due to a possible viral agent, it was observed that 30% of guppy farms in Agrotechnology Parks were affected compared with 70% of farms in the non-agrotechnology areas.

Table 2 summarizes some disease control practices used in 34 marine farms surveyed in 1992. Generally farmers did not separate newly introduced stock from those already present on the farms. However the practice of administering bath antibiotic and parasiticide treatments to new stock was quite common.

CONSTRAINTS

A problem not peculiar to Singapore, but perhaps common to most aquaculture industries in the region, relates to the nature of fish diseases found here and our present level of understanding of them. Conventionally, health programs such as those for domestic animals or farmed salmonids, focus on specific diseases of attributable economic impact, and the use of standard diagnostic procedures along with preventive and control programs. They involve the setting of specific targets *e.g.*, "reducing the prevalence of disease 'A' by 'X' % within a period," and are possible by virtue of the relatively comprehensive amount of information available. Commercial diagnostic kits and vaccines are also available, which encourages disease control. Given the fact that most fish diseases occurring in the region are still poorly understood, these are not yet possible here. Many diseases are considered as "syndromes" involving complex host-environment-pathogen interactions. Moreover, in ornamental fish culture, the numerous species involved could potentially be affected by an equally wide variety of pathogens. Even with the emergence of an increasing number of new pathogens, notably the viruses, it may be difficult to justify the development of new cell lines, diagnostic kits and vaccines to provide for the large number of species groups involved. In most cases, the respective sizes of these species groups, in terms of total market value, are still relatively insignificant.

As previously mentioned, aquaculture is becoming increasingly intensive in Singapore due to the limited availability of land and sea space. Only 1,700 ha of land has been designated for farming, and of that, slightly more than 10% is currently used by over 100 aquaculture farms. Consequently risks from disease will also increase, and must be addressed through improved health management programs.

Another constraint presently encountered is that health records such as mortality figures are often poorly kept by farmers. This makes assessment of mortality patterns during disease outbreaks difficult.

HEALTH PROBLEMS IN SINGAPOREAN AQUACULTURE

Quite loosely, we can classify the diseases seen in Singapore as being either (a) those associated with opportunistic pathogens or (b) those caused by specific agents.

Diseases associated with opportunistic pathogens

These are persistent, endemic diseases generally precipitated by some primary insult to the host. They become more significant with increased farming intensity and poor husbandry. They are often presented as disease syndromes involving fluctuating or poor water quality, unsuitable soil conditions, poor nutrition and usually concomitant infections by multiple opportunistic agents. Nonetheless they constitute a larger overall economic impact in comparison with problems caused by specific agents. Conditions in this category which are commonly seen include gill hyperplasia, erosive and ulcerative dermatitis of the fins and general body surface, bacterial septicemia, ectoparasitic infections, endoparasitic encystments and parasitic enteritis.

Common parasites found among ornamental fish cases include members of the genera *Ichthyobodo, Dactylogyrus, Gyrodactylus, Trichodina, Ichthyophthirius* and *Piscinoodinium*. Those commonly found affecting marine foodfish include *Benedenia, Cryptocaryon,* and *Trichodina.*

The bacteria isolated from many disease cases were mixed, and were usually considered to be insignificant. Tables 3 and 4 summarize the bacteria recently isolated from ornamental and marine fish by the Central Veterinary Laboratory and the Fish Health/Environment Unit (PPD), respectively. The interpretation of bacteriological and parasitological findings can be baffling; as a result much attention has been given to case histories, post-mortem findings and histopathology.

Recent observations suggest that there may be specific strains of bacteria, such as some *Vibrio harveyi* which are pathogenic to seabass, indicating some promise for the development of bacterial vaccines for marine foodfish. It is also likely that certain strains of *Aeromonas* may be more pathogenic than others to freshwater ornamental fishes, however, this needs further investigation. Likewise, outbreaks which have been attributed to certain parasites such as *Argulus, Lernaea* and *Hexamita* have been observed. Because these parasitic infections respond to specific treatments (*e.g.*, metronidazole to treat *Hexamita* infections), they are thought to be primary diseases under certain circumstances.

Diseases caused by specific agents

These include the new diseases and are characterized by some or all of the following:

- they affect apparently healthy stocks in relatively well-managed farms
- they cause mortalities of epizootic proportions
- many farms are affected concurrently
- they present with specific, often pathognomonic gross or histopathological signs or lesions
- both young and growout stocks are affected.

In Singapore, viral diseases have dominated this category. Some of them are novel diseases and have caused epizootics in specific species and age groups. The following have been observed:

Marine foodfish

- Lymphocystis of seabass (Chao, 1984)
- Rabbitfish viral hepatopathy (unpubl. obs.)
- Red grouper reovirus infection (Chew-Lim *et al.*, 1992)
- Viral nervous necrosis (VNN) of brownspotted grouper (Spinning Grouper Disease) (Chua *et al.*, 1993)
- VNN of seabass (unpubl. obs.)
- Bent body syndrome/swimbladder syndrome of brownspotted grouper (Chua *et al.*, 1993)
- Sleepy grouper disease of brownspotted grouper (Chua *et al.*, 1994)

Freshwater ornamental fish

- An infectious disease affecting angelfish in 1988 (Lim S.J., pers. comm.).
- An infectious disease affecting guppies in 1993 (unpubl. obs.).

DRUG AVAILABILITY AND USAGE

Importers of drugs for aquaculture use must be licensed under the *Poisons Act*, which is administered by the Ministry of Health (MOH). As a condition before purchase of drugs, farmers must produce their farm license; details of the purchase, such as quantity and date of purchase are recorded and maintained by the supplier. These records are checked periodically by MOH officials. For importation of new drugs or chemicals, advice from the PPD is first sought.

Drugs and chemicals are commonly used during husbandry operations such as grading and sexing, or in between transfer into new tanks/nets. Many farmers carry out preventive treatment of newly introduced stocks. Exporters of ornamental fish sometimes add antibiotics and antiseptics to the packing water used for transportation. In such prophylactic procedures, broad spectrum medications are used, targeting bacteria and ectoparasites. Table 5 lists the commonly used drugs and chemicals. Both generic and nongeneric drugs are used.

Culture-sensitivity tests conducted for bacteria isolated from marine fish over the last two years indicate consistent *in-vitro* resistance of many *Vibrio* isolates to tetracycline and oxytetracyline. There was little or no resistance to the other antibiotics tested (amoxycillin, chloramphenicol, erythromycin, trimethoprim and furazolidone).

GOVERNMENT APPROACH TO HEALTH MANAGEMENT

The primary aims of aquatic health management in Singapore are to reduce the prevalence of diseases due to opportunistic pathogens and to minimize the impact of introduced diseases. Under the *Fisheries Act*, the PPD may enforce legislative controls to prevent the introduction and spread of diseases when deemed necessary. Import regulations require the licensing of all fish importers and the declaration of every consignment of imported fish.

A Farm Accreditation Scheme was established in 1989 to cater to the export trade. Under the scheme, the PPD routinely inspects farms to ensure that minimum standards of disease control, such as quarantine facilities and procedures, are in place. Fish are routinely sampled at the point of export for clinical and laboratory examinations.

A fish health surveillance program which will involve the regular monitoring of a wide spectrum of farms is being studied. In particular, certain farms (sentinel farms) may be selected in rotation for closer monitoring of disease and mortality patterns, and of on-farm disease control systems. Particular batches of fry may be followed through to harvest to help further understand disease trends. The emphasis would also be on strategic sampling of stock for routine examination. High risk groups, such as fry, would be identified so that their monitoring may be concentrated during high risk periods (*e.g.*, during dry spells or monsoons). The push for on-farm quarantine in inland farms (*e.g.*, ornamental fish farms) and prestocking treatment of introduced marine fish will continue. Standardization of such procedures will be considered.

Improved facilities for clinical observation of fish, particularly in the study of new diseases, are being planned. Information transfer is also carried out during visits to farms, and seminars are held periodically.

The government's approach to research and development in fish disease is to link up with academic institutions and the private sector. Typically, research areas are identified by the government. Thereafter, the basic research component, such as molecular work, may be taken up by academic institutions, and subsequent development of product lines (*e.g.*, diagnostic kits, vaccines) by private companies. Currently, the PPD is collaborating with Ngee Ann Polytechnic in the serotyping and genetic fingerprinting of *Vibrio* isolates from marine foodfish. Another project between the PPD, the NUS and a private company dealt with the development of fish cell lines and the isolation and identification of certain fish viruses.

A major spin-off from the establishment of Agrotechnology Parks by the PPD is that health management can be better facilitated. In these parks, farms can be kept reasonably isolated and quarantined should the need arise. The quality of effluent can also be monitored and controlled more easily. Within the farms, water recycling systems are being developed. It is also more feasible to put into place systems which isolate culture units, provide for more efficient treatment procedures and segregate batches of stock in these farms. This Agrotechnology Park concept will continue to shape the nature of fish farming in Singapore.

CONCLUSIONS

While Singapore's production base remains modest by virtue of its limited area for agricultural development, aquatic animal health management is nevertheless a high priority. As the approach adopted is principally that of "high-tech farming" or agrotechnology, the justification for development and production of high-value stock with good health status is even greater. Increased health vigilance is anticipated with the expected increase in aquaculture trade between Singapore and her neighbors in the Asia-Pacific.

ACKNOWLEDGMENTS

I thank the Director of Primary Production for permission to present this paper, and Miss Julie Goh for her assistance in preparing the manuscript.

REFERENCES

Chao, T.M., 1984. Studies on the transmissibility of lymphocystis disease occurring in seabass, *Lates calcarifer* (Bloch). Singapore J. Pri. Ind. 12:11-16.

Chew-Lim, M., G.H. Ngoh, S.Y. Chong, F.H.C. Chua, J.L.C. Howe and E.W. Lim, 1992. Description of a virus isolated from the grouper *Plectropomus maculatus* (L). J. Aquat. Anim. Health, 4: 222-226.

Chua F.H.C, J.J Loo, J.Y. Wee and M. Ng, 1993. Findings from a fish disease survey: an overview of the marine fish disease situation in Singapore. Singapore J. Pri. Ind. 21: 26-37.

Chua F.H.C., M.L. Ng, K.L. Ng, J.J. Loo, and J.Y. Wee, 1994. Investigation of outbreaks of a novel disease, `Sleepy Grouper Disease', affecting the brown-spotted grouper, *Epinephelus tauvina*, Forskal. J. Fish Dis. 17: 417-427.

Table 1. Disease control practices observed among eight Agrotechnology Park ornamental fish farms surveyed in 1992.

1. Drying and disinfection of ponds and tanks between batches
2. Partial or complete water change of spawner tanks periodically
3. Bath treatment with salt, acriflavine or other disinfectants (preventive and therapeutic) during an outbreak or during other husbandry operations
4. Feeding of medicated diets (*e.g.*, feed mixed with metronidazole)
5. Water change during an outbreak

Table 2. Disease control practices used by 34 marine netcage farms surveyed in 1992, in decreasing order of frequency.

1. Treatment during the course of a disease outbreak
2. Regular removal of floating fish carcasses
3. Sanitation of newly-introduced stock
4. Reduction of stocking densities during outbreaks
5. Stocking of carrion feeders (*e.g.*, lobsters) in finfish nets
6. Isolation of sick fish (in "hospital" nets)
7. Treatment of fish when feeding response is reduced
8. Aeration during periods of stress (*e.g.*, during plankton blooms)

Table 3. Summary of bacteria isolated from cases involving diseased ornamental fish from January 1995 to March 1995.

Date	Fish species	Bacteria isolated
03/95	neon tetra	*Aeromonas sobria*
	swordtail	*A. sobria*, *Vibrio* sp.
	guppy	*Acinetobacter* sp., *Alcaligenes* sp., *Flavobacterium* sp., non-01 *V. cholerae*, *A. hydrophila*, *V. cholerae*, non-haemolytic *Escherichia coli*
02/95	giant danio	*A. sobria*, *Vibrio* sp.
	guppy	*Acinetobacter* sp., *Alcaligenes* sp., non-01 *V. cholerae*, *A . hydrophila*, *Vibrio* sp., *A. sobria*, *A. caviae*
01/95	tiger barb	*Aeromonas* sp.
12/94	bitterling	*A. hydrophila*, *A. sobria*
	silver dollar	*Plesiomonas shigelloides*
	tiger barb	*A. hydrophila*
	cichlid	*A. sobria*
11/94	tiger barb	*Aeromonas* sp.
10/94	kissing gourami	*A. hydrophila*, *P. shigelloides*
	harlequin	*A. sobria*
09/94	platy	*A. hydrophila*
08/94	goldfish	*Aeromonas* sp.
	Malawi cichlid	*Edwardsiella tarda*
	neon tetra	*Edwardsiella* sp.
	catfish	*A. hydrophila*
	angel	*Alcaligenes* sp.
06/94	glowlight tetra	*E. tarda*
	swordfish	*A. sobria*
02/94	neon tetra	*E. tarda*
	black molly	*A. sobria*
01/94	dwarf gourami	*A. sobria*, *P. shigelloides*

Table 4. Bacteria isolated from cases involving diseased marine fish from April 1993 to January 1995

Date	Fish species	Bacteria isolated
01/95	seabass	*Vibrio* sp., *A. salmonicida* (mas/achro)
11/94	greasy grouper	*Vibrio* sp., *Aeromonas* sp.
10/94	seabass	*V. parahaemolyticus*, *Vibrio* sp.
08/94	seabass	*Vibrio* sp.
07/94	polkadot grouper	*Vibrio* sp., *A. salmonicida* (mas/achro)
	seabass	*V. vulnificus*
06/94	tiger shrimp	*V. alginolyticus*
05/94	seabass	*V. vulnificus*
04/94	tiger shrimp	*V. alginolyticus*, *Aeromonas* sp.
	tiger shrimp	A. *hydrophila*, *A. caviae*, *Aeromonas* sp.
12/93	seabass	*V. harveyi*
	greasy grouper	*V. damsela*
	seabass	*Vibrio* sp.
11/93	seabass	*Streptococcus* sp., *V. harveyi*
10/93	red tilapia	*Vibrio* sp., *Pseudomonas* sp.
08/93	pompano	*V. harveyi*, *Vibrio* sp.
	red grouper	*V. vulnificus*, *Pasterella haemolytica*
07/93	seabass	*V. vulnificus*, *V. harveyi*, *Vibrio* sp.
06/93	seabass	*V. vulnificus*, *V. harveyi*, *Vibrio* sp., *Plesiomonas shigelloides*
05/93	seabass	*V. vulnificus*
05/93	seabass	*V. vulnificus*, *Vibrio* sp. *A. hydrophila*

Table 5. Therapeutic agents commonly used in Singaporean aquaculture.

- **Antibiotics and antibacterials**

Trimethoprim
Furazolidone
Nitrofurazone
Monofuran
Tetracyclines
Neomycin
Erythromycin
Chloramphenicol
Sulphonamides, Sulfathiazole, Sulfamonomethoxine
(*e.g.*, "Daimeton")

- **Antiseptics and parasiticides**

Formalin
Malachite green
Methylene blue
Copper sulphate (*e.g.*, "Protocide")
Acriflavine
Potassium permagenate
Mercurochrome
Organophosphates (e.g., "Dipterex", "Masoten")
Flubendazole
Metronidazole
Benzylkomium chloride (BKC)
Hydrogen peroxide
Chlorine dioxide ("Aquaplus")
Iodophors

- **Others**

Sodium chloride (for freshwater fish)
Fresh water (for marine fish)

Shrimp Farming in Sri Lanka: Health Management and Environmental Considerations

P.K.M. Wijegoonawardena and P.P.G.S.N. Siriwardena

National Aquatic Resources Agency, Colombo 15, Sri Lanka

Wijegoonawardena, P.K.M., and P.P.G.S.N. Siriwardena, 1996. Shrimp farming in Sri Lanka: health management and environmental considerations. *In* Health Management in Asian Aquaculture. Proceedings of the Regional Expert Consultation on Aquaculture Health Management in Asia and the Pacific. R.P. Subasinghe, J.R. Arthur & M. Shariff (eds.), p. 127-139. FAO Fisheries Technical Paper No. 360, Rome, FAO. 142 p.

Abstract

Aquaculture is a recent activity in Sri Lanka. However, farming of penaeid shrimp, *Penaeus monodon*, has been a successful and lucrative venture until major disease outbreaks occurred in late 1980s. Although the main cause for these outbreaks is considered to be the introduction of an exotic viral pathogen, uncontrolled proliferation of farm operations and related aquatic environmental implications appear to have made a direct contribution. This paper describes the status of shrimp farming in Sri Lanka, with special reference to health and environmental issues which prevail in the sector.

INTRODUCTION

Sri Lanka does not have a tradition of aquaculture practice; there was virtually no aquaculture until the beginning of the 1980s. Interest in brackishwater aquaculture was generated with the establishment of the Brackish Water Fisheries Station in Negombo as a research and demonstration center. In 1968, the Ceylon Fisheries Corporation constructed ponds on a 20 ha tidal flat, but abandoned them in 1970 after the dikes were washed away. In 1977, a small shrimp farm covering 0.7 ha began operations in the shallow edge of the Batticoloa Lagoon, but had to be abandoned without further production due to civil unrest in the country.

Until the mid-1960s, the shrimp fishery was entirely a lagoon and estuarine by-catch fishery. With the demand in the international market continuously rising, trawling at sea for shrimp developed rapidly and production from this capture fishery reached a peak of around 4 000 metric tonnes (mt) in the 1980s (Siriwardena, 1989 - unpubl. doc.). With the problems associated with civil disturbances in the north and east, fishing there was suspended, while trawling in the western and northwestern coasts showed signs of depleting the shrimp resources. Given the importance of increasing shrimp production and the availability of unpolluted brackish water and land suitable for shrimp culture that is essentially unsuitable for agriculture purposes, the Government of Sri Lanka promoted private sector investments in shrimp farming by offering various incentives. In response to the high demand in the world market and ever-increasing export prices, local and multinational companies and small-scale entrepreneurs embarked on shrimp farming. Over the last decade there has been a rapid development of shrimp farms in the northwestern province of Sri Lanka, but this has been accompanied by environmental and disease problems which have led to the closing of some farms and loss of productivity in others.

Shrimp culture operations are entirely in the hands of private sector entrepreneurs who use local labor and, partly, expatriate technicians. Large areas of land adjacent to lagoons have been utilized for shrimp culture, which developed rapidly on the western and northwestern coasts in the mid-1980s. Operations along the east coast ceased after the 1983 civil disturbances. To date, more than 300 entrepreneurs have undertaken shrimp farming, and a total land area of more than 2 000 ha has been utilized in the northwestern and western coastal areas for the development of shrimp farms.

Apart from commercial-scale pond farming of shrimps, experimental shrimp culture in pens was attempted in 1986. Trials were carried out in Chilaw and Negombo lagoons and production levels of 400 to 600 kg/ha/crop of *Penaeus monodon* and *P. merguiensis* were achieved on a semi-intensive scale of practice.

COASTAL RESOURCES FOR SHRIMP CULTURE

The coast of Sri Lanka is endowed with large areas of saltwater lagoons, estuaries, mangroves, and coastal mud flats. Thirteen of the 24 administrative districts in Sri Lanka have maritime boundaries and the development of coastal aquaculture is therefore restricted to these districts. The extent of water area suitable for coastal aquaculture in pens is estimated to be around 12 000 ha (Weerakoon and Siriwardena, 1988). The water area for the development of cage culture in brackish water is negligible (Siriwardena, 1982), since areas deeper than 2 m are restricted to the navigational paths of most lagoons. There is about 6 000 ha of land suitable for shrimp farming in Sri Lanka, of which around 57% and 20% are situated along the northeast and northwestern coastal areas (Samaranayake, 1986). The National Aquatic Resources Agency (NARA) of Sri Lanka has identified land areas of 1 200 ha and 250 ha as suitable sites for shrimp farming in the northwestern and southern coastal areas, respectively (NARA, 1989).

SHRIMP CULTURE PRACTICES AND PRODUCTION

Share of shrimp aquaculture in the fisheries sector

The fishery sector is an important part of Sri Lanka's economy and contributed around 2.2% of GDP in 1993. Fish is the most important source of animal protein consumed in the country, about 60% of the population depending solely on fish for their protein requirement, and accounting for nearly 65% of the total animal protein consumed. The steady growth in local production and increased fish imports led to a rapid increase in *per caput* supplies from 10.7 kg in 1977 to 16.9 in 1989. The bulk of this production came from the coastal marine fishery where in some areas there has been over-exploitation. Inland fish production is heavily dependent on capture fisheries in the reservoirs and tanks. The marine offshore/deepsea fishery is at present more or less confined to the western and southern coasts and has shown a steady growth from an estimated 3 000 mt in 1978 to 33 000 mt in 1993. The sector employs around 150 000 in fishing, supporting a fishing population of about 600 000. In addition, around 13 500 fishermen were involved in inland fishing and another 50 000 are estimated to be engaged in supporting activities, including fish processing and marketing. Employment in fishing represents about 1.75% of total employment, and about 4% of agricultural employment. When ancillary activities are also considered, the respective shares amount to about 2.5 and 5%.

The contribution of shrimp aquaculture production to the total fish production increased from 0.06% (100 mt) to 0.7% (1200 mt) between 1985 and 1993. It has created more than 4 000 jobs, or around 2.0% of fisheries sector employment.

Growth of the shrimp culture industry

Interest in brackishwater shrimp culture in Sri Lanka started in 1977, but the commercial production first entered the market in 1984. In 1984, only one multinational company was producing brackishwater shrimp for the export market. The number of operational establishments had risen to 52 in 1989 and by the end of 1994, the number of authorized farms had increased to 250. The total number of farms including unauthorized farms far exceeds this number.

The water area under culture (excluding bunds) increased from 3 to 12 ha from 1984 to 1986). Due to the expansion of semi-intensive culture practice, the water area under culture increased rapidly in 1997 to 125 ha. By 1989 the area under culture had increased to 243 ha; since then, no proper records have been maintained. The area allocated, however, for authorized shrimp farms amounted to 1 400 ha at the end of 1994. Once again, the total area developed, including unauthorized farms, far exceeds this total and is distributed along the northwestern coastal belt. The estimated employment generated through shrimp farming was over 4 000 at the end of 1994.

Shrimp culture practices

Thirty-one species of shrimp have been recorded from Sri Lankan waters. Of these, only the penaeid shrimps have commercial value (de Bruin, 1970). The operational shrimp farms in Sri Lanka practice monostock monoculture of *Penaeus monodon* (black tiger shrimp), a species which has been selected for culture due to its large size, fast growth, high price, high market potential and the availability of technology (Siriwardena, 1990 - unpubl. doc.).

The shrimp farming industry in Sri Lanka is dependent entirely on hatchery-bred post-larvae. The present requirement of hatchery-bred post-larvae is around 450 million per year, while the present hatchery production is very inadequate, being around 200 million post-larvae per year (Siriwardena, 1990 - unpubl. doc.).

Culture practice in the mid-1980s was largely intensive; however, this trend changed after the disease outbreak in 1988/89. The farmers who were committed to intensive culture used reduced stocking densities and two crops per year, thereby allowing time for treatment. In late 1989, only seven of 52 establishments were practicing intensive culture (Siriwardena, 1990 - unpubl. doc.). The production rates for intensive and semi-intensive culture ranged between 10-15 and 6-8 mt/ha/yr, respectively. The present trend is to practice semi-intensive culture employing a stocking rate of 15-20 post-larvae/m^2 with aeration and artificial feeding.

Of the three sizes of shrimp farm, the small sector comprises the largest number. These farms are viable, and their numbers are expected to increase because of government policy not to permit the existing large farms to become larger and not to approve any more large-scale shrimp farms (Joseph, 1993).

Shrimp culture production

Government plans

Shrimp and finfish culture is being accorded high priority in the National Fisheries Development Plan for 1995 to 2000. Production of cultured shrimp is projected to increase to 5 500 mt, from the present production of 1 200 mt (wet weight).

Shrimp culture production and exports

Shrimp comprises the largest quantity of any export product in the fisheries sector and earns the highest foreign exchange. The percentage contribution of shrimp to the total value of all aquatic products exported varied between 48.5% and 70.3% between 1985 to 1992. The decline in production in relation to the water area under culture can be attributed to the progressive deterioration of the pond environment due to intensive culture and conversion to semi-intensive culture. The contribution of cultured shrimp to total shrimp exports in terms of quantity and value increased from 1985 to 1993, as a capture fishery did not take place in the north and east due to civil disturbances.

HEALTH AND ENVIRONMENTAL CONSIDERATIONS

Health and environmental concerns have become major issues in many countries where aquaculture has developed to a commercial scale. These concerns include disease outbreaks, effluent discharge, use of antibiotics, habitat change and damage, and the adverse effects of pollutants on the aquatic environment.

Problems related to health management

The first disease outbreak which significantly affected production occurred during 1988 to 1989. The following disease signs were observed (Jayasinghe, 1995):

- microfouling on shells
- reduced frequency of molting
- reduced feeding
- black gills
- soft shell conditions
- tail rot
- black spot
- red/brown deposits on the belly and red coloration

A drop in production from 5.3 to 1.9 mt/ha/cycle causing an average reduction of 64% in production and an estimated total loss of 186.62 Rs. million (US$ 4.44 million) was observed due to this disease outbreak (Jayasinghe, 1995).

No proper monitoring program for health management and disease has been undertaken since the inception of this industry. Hence, estimation of losses due to diseases in terms of reductions in quantity and value is difficult. Brown and black gill conditions and MBV virus were reported in grow-out facilities. Iron hydroxide deposits, hypertrophic changes and hemolytic infiltrations were found among gill lamellae, plus infestations by the ectocommensal protozoan *Zoothamnium* (Jayasinghe, 1995; Wijegoonawardena, unpubl. data).

Total bacterial counts of 2.4×10^4 to 9.0×10^5 and 5×10^2 to 8.8×10^3 were reported for cultured shrimp and grow-out pond water, respectively (Fonseka and Hettiarachchi, 1991). Of this, 19.4% and 5% were *Vibrio* and *Pseudomonas*, respectively. These bacterial counts, however, were made for post-harvest technological purposes and it is difficult to make any inferences on the influence of these bacterial levels during culture.

Luminescent vibriosis has been recently recognized as the most serious disease affecting hatchery production of *P. monodon*. Luminescent strains of *Vibrio* species have been implicated in outbreaks of shrimp hatchery bacterial diseases mainly in the northwestern coastal areas of Sri Lanka. The high levels of luminescent vibrios in the sea water of this area may be due to high organic matter, as according to Shilo and Yetinson (1979) these bacteria grow very well in eutrophic sea water. According to hatchery operators of this coastal area, water quality parameters such as temperature, salinity, pH, ammonia, nitrate and phosphate have no effect on the occurrence of luminescent vibrios. In other countries, two species, *Vibrios harveyi* and *V. splendidus,* have been identified, with *V. harveyi* being more dominant. Significant larval mortalities associated with luminescent vibriosis caused by *V. harveyi* and *V. splendidus* were reported from hatcheries raising *P. monodon* and *P. merguiensis* in Indonesia (Sunaryanto and Mariam, 1986) and in the Philippines (Baticados *et al.*, 1990). The strain(s) of *Vibrio* causing mortalities in hatcheries in Sri Lanka has yet to be determined. In Sri Lanka, vibriosis has been noted to cause up to 100% mortality of *P. monodon* from zoea and mysis to post-larval stages. This has aggravated the problem of production shortage in shrimp post-larvae, leading to the recent importation of *P. monodon* post-larvae into the country. Such importations without proper quarantine measures may cause serious disease outbreaks in the shrimp farming industry.

While most disease control programs emphasize the pathogen (*e.g.*, microbial diagnosis, vaccine development and chemotherapeutic treatment), one must not forget that poor water quality and inadequate nutrition are often basic determinants of disease outbreaks (Sindermann and Lightner, 1988). Questions have been raised as to whether the inter-connected lagoon complex between Chilaw and Puttalam can continue to sustain more than 1 200 ha of shrimp ponds. The canal that serves as the main source of brackish water for the farms is apparently polluted and the risk of disease outbreaks is increasing.

There is little information on the amounts of different chemotherapeutants used in the shrimp farming industry, but the increase in their use is causing concern, both from human health and the environmental aspects. The commonly use antibiotics are chloramphenicol, oxytetracycline, terramycin and nitrofuran.

Problems related to the environment

In brief, the problems related to the environment in the shrimp farming industry in Sri Lanka arose mainly from over-emphasis on high production, economic viability and foreign income generation without full consideration of the environmental impacts caused by over-crowding of farms. At the inception, there was no proper zoning plan to facilitate development of an environmentally sound industry. Lack of a proper zoning plan for the northwestern and western provincial coastal belts led to many social problems and destruction of ecologically sensitive areas, such as mangroves and mud flats. Moreover, this destruction has not yet been quantified in terms of ecological importance and value.

When approving a particular site, neither the industry as a whole nor the existing farms in the vicinity were taken into consideration. This led to the establishment of farms in a haphazard manner so that the outlets and inlets of individual farms were located in close proximity, inevitably allowing one to obtain the discharged water from the other and leading to health management problems.

The main water sources serving the present shrimp farming industry are Negombo Lagoon, Chilaw Lagoon, Puttalam Lagoon and the canal connecting these three lagoons, the Dutch Canal. According to the renewal rates and quality of water measured in early 1990, the canal, extending from Chilaw to Mundal has the capacity to supply water to a culture area of 242 ha (Siriwardena and Dayaratne, 1990 - unpubl. doc.). Siriwardena and Dayaratne (1990 - unpubl. doc.), recommended improvements to the Dutch Canal in order to enhance the water renewal rate if the shrimp farming industry was to be expanded in the northwestern coastal belt. However, since 1990, the number of shrimp farms has increased rapidly in this area, exceeding the culture area of 242 ha without any improvements in water renewal rates of the Dutch Canal. Many water quality parameters of the Dutch Canal changed due to effluent discharge from shrimp farms (Table 1). Increased suspended solids may cause a further reduction in the volume and renewal rates of water in the Dutch Canal. If the dissolved oxygen depletes, the increased levels of total sulfides may cause environmental stress leading to fish kills. Such fish kills were reported in the Dutch Canal in early 1995. Moreover, reduced growth rates, lower production and disease problems in shrimp may be attributed to the stress caused by the polluted water source.

The present trend in the northwestern coastal belt is to establish shrimp farms in high saline areas (beyond Mundal up to Puttalam and Kalpitiya) due to unavailability of sites in low saline areas (between Chilaw and Mundal) for development. With the intensification of shrimp farming in high saline areas, the use of fresh water by installing tube wells has increased. Establishment of shrimp farms in high saline areas without assessing the adequacy and status of the water table may result in its depletion, leading to land depression. In addition, too much dependence on underground water may result in a much lower water table, which, in turn, may lead to serious competition for fresh water between aquaculture, agriculture, industry and domestic users and to drinking water becoming salty.

The establishment of many small-scale farms without proper environmental assessment led to the destruction of buffer zones between farms and between water fronts and farms and to interference with the irrigation and drainage systems. The latter causes annual flooding in the areas where shrimp farms are developed.

RECOMMENDATIONS

Disease considerations

There is a need for a regular monitoring program to determine the health status and to recommend mitigatory measures to prevent possible disease outbreaks. Lack of proper monitoring makes it difficult to estimate economic losses caused by disease. A monitoring program should emphasize the following:

- Monitoring of each farm for changes in the main environmental parameters, such as dissolved oxygen deficit, accumulation of toxic nitrogen compounds, toxic algae, external pollution etc. This will enable collective measures to be taken to remedy any adverse changes before an outbreak of disease occurs. The monitoring may be done by farmers who are capable of doing so. NARA may monitor the small-scale farms and coordinate the entire program.
- Monitoring conditions that increase susceptibility to infectious diseases through stress or which cause acute effects.
- Monitoring the use and effects of chemotherapeutants. The use of antibiotics may result in some environmental changes, such as:
    - qualitative and quantitative changes in the bacterial flora
    - toxic effects on wild living organisms
    - development of antibiotic resistance in bacteria
    - transfer of antibiotic resistance to bacteria pathogenic to humans (Anon., 1991).

Due to the polluted nature of the main water sources used for shrimp farming, there is a great possibility for disease to spread throughout the area. Poor design and unplanned localization and siting of farms aggravate the risk, since infection through the water is a major source of disease transmission. Therefore, improvements such as dredging of the Dutch Canal and opening the sea out-falls are recommended to improve the water exchange rates (Siriwardena and Dayaratne, 1990 - unpubl. doc.).

A risk analysis for health and disease management is recommended. The following areas should be examined:

- Localization of the farm, water resources and water quality
- The intake of live material at the farm
- Farming strategy, including the control of the production cycle and links with other farms
- Hygienic conditions
- Management routines
- Monitoring and control routines

Environmental considerations

There is a need for the Government of Sri Lanka to recognize the important role of aquaculture in its national economy, so that proper laws and regulations beneficial to its controlled expansion are established. One of the important strategies for the development of shrimp aquaculture is the adoption of transparent policies to avoid the social conflicts and adverse ecological impacts which are currently occurring. Zoning plans, based on scientific investigations and analysis of physical, biological, social and economic information of the areas, especially in the new areas to be developed, would provide a rational approach to the allocation of land and water areas and the development of improved pond designs.

Freshwater resource use must be rationalized. Shrimp farms which are dependent upon fresh water for dilution should not be approved without knowing the availability and adequacy of the natural water table for competitive use without affecting the environment. Prior to approving further shrimp farms in the high saline areas, the following questions should be adequately addressed:

- What is the water quantity needed to dilute high saline water in the shrimp farms?
- Can the natural water table meet the requirement of fresh water for the shrimp farms?
- Has the water table been depleted or is it on the verge of depletion in the areas where shrimp farms depend on fresh water for the dilution of high saline water?
- What quantity of fresh water could be allowed for use to adjust the salinity in shrimp farms without creating user conflicts?

Besides employing the above measures to conserve water resources, construction of water recycling facilities in several demonstration farms to save water and reduce the impact on the environment should be encouraged. Placing a subscribed charge on water pumped from underground would create a water conservation consciousness among farmers.

A "switching over" strategy should be adopted to change the species under culture to a relatively high salinity tolerant species. For example, relatively high salinity tolerant white prawn (*Penaeus indicus*) could be cultured in high saline areas instead of black tiger prawn, which prefers low saline conditions. This may prevent or minimize the use of fresh water to dilute high saline water in shrimp farms. Moreover, it is generally recognized that high saline species are relatively more disease tolerant than are low saline species. When adopting the "switching over" strategy, the species should be carefully selected, taking into consideration the market trends.

Has cultured shrimp production increased in relation to the increased culture area enough to justify approving a vast number of shrimp farms in the northwestern provincial coastal belt? The production figures show that it has not increased in relation to the increase in culture area. This cannot be totally attributed to the switch over from intensive to semi-intensive practice. It is partly because of the inability of hatchery production to meet the demand and partly because of the slowdown of growth rate of shrimp experienced, probably due to the polluted nature of the water source. Hence, to achieve effective culture, approval of land has to be scaled down, giving due consideration to actual hatchery production and pollution status of the water source.

To reduce the impact of shrimp farming, the "cluster farm" concept may be adopted. Almost all small shrimp farms are being operated without any mitigatory measures, such as the use of oxidation and sedimentation tanks to reduce the hazardous impact on the water source. Depending on the size of the land, the number of such small farms (not exceeding one ha in extent) in a particular area could vary. The "cluster farm" concept may be adopted in the following manner:

- Identify areas where small farms are in close proximity.
- Group small farms with a total extent not exceeding 2-3 ha to form a small unit, bringing in the concept of "cluster farms".
- Such clusters would have the following common structures, if desired, and if location allows such amalgamation:
    - common inlet
    - common outlet
    - common pumping station
    - common sedimentation and oxidation tank

This approach would minimize the problem of locating inlets and outlets at close proximity and reduce the sediment and nutrient loading in the water source by utilizing organized common sedimentation and oxidation tanks.

Adoption of a system to monitor environmental licenses to prevent establishment of unauthorized farms without proper environmental assessments is important.

There was no frequent and proper dialogue between the researcher, the developer and the extensionist. Without proper dialogue, it is difficult to obtain the feedback needed for formulating research to address the problems in the industry. Hence, to establish a proper dialogue between these groups via regular meetings of their representatives is recommended.

REFERENCES[2]

Anon., 1991. Fish health management in Asia-Pacific. Report on a regional study and workshop on fish disease and fish health management. Asian Development Bank and Network of Aquaculture Centres in Asia-Pacific. ADB Agric. Dep. Rep. Ser. No. 1.

Baticados, M.C.L., C.R. Lavilla-Pitogo, E.R. Cruz-Lacierda, L.D. de la Pena, and N. Sunaz, 1990. Studies on the chemical control of luminous bacteria *Vibrio harveyi* and *V. splendidus* isolated from diseased *Penaeus monodon* larvae and rearing water. Dis. Aquat. Org. 9: 133-139.

de Bruin, 1970. The distribution of penaeid prawns in Ceylon water. Bull. Fish. Res. Stn. Cey. 21: 75-86.

Fonseka, T.S.G., and K. Hettiarachchi, 1991. Incidence of some pathogenic organisms in cultured shrimp (*Penaeus monodon*) collected from Chilaw. Vidyodaya J. Sci. 3: 23-33.

Jayasinghe, J.M.P.K., 1995. Sri Lanka country report, 357-376 p. *In* Report on a Regional Study and Workshop on the Environmental Assessment and Management of Aquaculture Development (TCP/RAS/2253). NACA Environment and aquaculture Development Series No. 1. Network of Aquaculture Centres in Asia-Pacific, Bangkok, Thailand, 492 p.

Joeseph, L., 1993. Coastal fisheries and brackish water aquaculture in Sri Lanka. Coastal Resources Management Project of the University of Rhode Island, United States Agency for International Development, 45 p.

NARA, 1989. Identification of suitable sites for shrimp farming: Phase I. North western coastal belt. A consultancy report for Export Development Board and National Development Bank. National Aquatic Resources Agency (NARA), Colombo, Sri Lanka, 291 p.

Samaranayake, R.A.D.B., 1986. Status and prospects for brackishwater aquaculture in Sri Lanka. J. Inl. Fish. (Sri Lanka), 3: 88-90.

[2] Editor's note: As there exists little published data on Sri Lankan shrimp culture, citations for the unpublished documents of Joseph (1993), NARA (1989), Siriwardena (1989, 1990), Sirwardena and Dayaratne (1990) and Weerakoon and Siriwardena (1988) are included as an aid to future workers.

Shilo, M. and T. Yetinson, 1979. Physiological characteristics underlying the distribution patterns of luminous bacteria in the Mediterranean Sea and the Gulf of Elat. Appl. Environ. Microbiol. 38: 577-584.

Sindermann, C.J., and D.V. Lightner, 1988. Disease diagnosis and control in North American marine aquaculture. Developments in Aquaculture and Fisheries Science. 17, Elsevier Science Publishing Co., Amsterdam.

Siriwardena, P.P.G.S.N., 1989. Survey on Aquaculture (Phase 2), Sri Lanka. Unpublished document prepared for the Asian Productivity Organization, Tokyo, Japan.

Siriwardena, P.P.G.S.N., 1990. Aquaculture in Sri Lanka's coastal environment: activities, contributions, conflicts and projections. Unpublished document prepared for the Coastal Resources Management Project, Department of Coast Conservation, Sri Lanka.

Siriwardena, P.P.G.S.N. and P.W.C. Dayaratne, 1990. Report of the technical committee on Dutch Canal rehabilitation for shrimp farming. Unpublished report prepared for the National Aquatic Resources Agency (NARA), Colombo 15, Sri Lanka.

Sunaryanto, A. and A. Mariam, 1986. Occurrence of a pathogenic bacteria causing luminescence in penaeid larvae in Indonesian hatcheries. Bull. Brackishw. Aquacult. Developm. Cent. 8: 64-70.

Weerakoon, D.E.M., and P.P.G.S.N. Siriwardena, 1988. Final report on cage and pen culture roject to the Internatinal Development Research Cernre, Canada. Inland Fisheries Division, Ministery of Fisheries, Colombo, Sri Lanka, 80 p.

Table 1. Changes in some water quality parameters in the Dutch Canal. [1]

Parameter	1983	1987	1992
pH	4.8 - 6.0	5.3 - 7.3	7.2 - 8.8
Salinity	0.0 - 26.0	0.0 - 33.0	0.0 - 33.0
Phosphate (ppm)	0.02 - 0.50	0.05 - 0.85	0.03 - 2.00
Nitrate (ppm)	0.01 - 1.60	0.01 - 2.70	0.02 - 3.40
Nitrite (ppm)	NA [2]	0.02 - 0.32	0.20 - 1.40
Sulfide (ppm)	NA	0.02 - 0.61	0.20 - 1.80
Temperature (°C)	23.0 - 32.0	23.0 - 32.0	25.0 - 32.0
Turbidity (NTU)	5.0 - 19.0	7.0 - 23.0	5.0 - 98.0
Total suspended solids (mg/L)	10 - 22	30 - 160	50 - 300

[1] Source: Jayasinghe (1995).
[2] NA = not available.

Quarantine Practices used in Papua New Guinea for Introductions and Transfers of Live Fish

Ursula M. Kolkolo

Department of Fisheries, P.O. Box 165, Konedobu,
National Capital District, Papua New Guinea

Kolkolo, U.M. 1996. Quarantine practices used in Papua New Guinea for introductions and transfers of live fish.. *In* Health Management in Asian Aquaculture. Proceedings of the Regional Expert Consultation on Aquaculture Health Management in Asia and the Pacific. R.P. Subasinghe, J.R. Arthur & M. Shariff (eds.), p. 140-142. FAO Fisheries Technical Paper No. 360, Rome, FAO. 142 p.

Abstract

The FISHAID Project (Fisheries Improvement through Stocking High Altitudes for Inland Fisheries Development) in Papua New Guinea (PNG) has been involved in the identification of appropriate edible cold- and warmwater fish species from Nepal, India, Malaysia and South America for introduction into PNG waters at higher and lower elevations, respectively. The drainage basin where this stocking program is being carried out is the Sepik and Ramu River Basin in northern PNG. This paper outlines the quarantine procedures that have been practiced in PNG during the introduction of six exotic species (chocolate mahseer, *Accrocephalus hexagonolepis*; snow trout, *Schizothorax richardsonii*; golden mahseer, *Tor putitora*; tawas, *Barbodes gonionotus*; pirapatinga, *Piaractus brachypomus*; and sabalo, *Prochilodus lineatus*); the transfer of common carp (*Cyprinus carpio*); and the importation of rainbow trout (*Oncorhynchus mykiss*) eggs.

BACKGROUND

The National Fisheries Authority of Papua New Guinea has two inland fisheries development programs. The first is the stocking of mid-altitude (about 1 300 m above sea level) streams with coldwater and warmwater fish species imported from overseas, and the second program is the pond culture of the common carp, *Cyprinus carpio*. Quarantine legislation dealing with the introduction of live fish requires that prior to introduction imported fish species be quarantined for three months. This legislation, known as the *Animal Disease Control Act*, also requires that Disease Certificates be issued for live fish imports. There is also the *Fauna Protection Act* which regulates the import and export of live organisms. Under this legislation, the National Fisheries Authority conducts, as much as possible, its inland fisheries development work. Within the private sector, there is a growing interest in information on commercial shrimp, milkfish, pearl-shell and oyster farming but, with the exception of two private farms culturing the rainbow trout,

Oncorhynchus mykiss, in the PNG highlands, no commercial operations have been established.

STOCKING OF INLAND WATERS

The current freshwater stocking program, called the FISHAID Project (Fisheries Improvement through Stocking High Altitudes for Inland Fisheries Development), stocks edible coldwater fish species imported from Nepal and India into higher coldwater streams and stocks warmwater species imported from Malaysia and South America into waters at lower elevations. The drainage basin where this stocking program is being carried out is the Sepik and Ramu River Basin in northern Papua New Guinea. The coldwater species imported from India and Nepal are the chocolate mahseer (*Accrocephalus hexagonolepis*), the snow trout (*Schizothorax richardsonii*) and the golden mahseer (*Tor putitora*). The warmwater species imported for stocking in the flood plains of the Sepik and Ramu River Basin are tawas (*Barbodes gonionotus*) and pirapatinga (*Piaractus brachypomus*, syn.: *Colossoma bidens*) from Malaysia and sabalo (*Prochilodus lineatus*) from Brazil.

Procedures for introductions and quarantine

The procedures followed for these introductions are basically the same and follow as much as possible a concept that is consistent with sustainable development:

- Prior to introduction, an initial environmental impact assessment is made of the site. This involves a description and, where possible, quantification of the natural freshwater fauna as well as a description of the biophysical characteristics of the natural streams.
- A list of potential species for introduction is drawn up and is assessed by a panel of international fishery scientists and aquaculturists. Here they consider the survival of the species concerned and its possible impact on native fauna.
- Live eggs are then imported from a fish culture station where disease-free certificates can be issued.
- Eggs are treated with iodine prior to shipment to Papua New Guinea.
- In Papua New Guinea, the fish are kept in quarantine for three months in the hope that if there are diseases present they will show within this period.
- After three months, live fish samples are sent to an overseas institution for disease testing So far Papua New Guinea has used the University of California at Davis and the Universiti Pertanian Malaysia. After the results are known, the fish are stocked.

TRANSFERS OF COMMON CARP FOR POND CULTURE

Common carp fingerlings are produced by a government hatchery and distributed throughout the country where there is interest in pond culture. There are no specific quarantine procedures to be followed; fingerlings are packed using the pond water of the government hatchery and air-freighted. Most of the fish are cultured in earthen ponds.

IMPORTATION OF RAINBOW TROUT

Private farms culturing rainbow trout import eggs from Australia. There is no quarantine involved, the reason being that because of historic ties between Australia and Papua New Guinea, PNG, under the Animal Disease and Control Act, can only import live fish eggs, specifically trout from Australia only. Furthermore, as Australia provides a disease-free certificate, the three month quarantine period is waived.

CURRENT CONSTRAINTS

There are significant inconsistencies and constraints within Papua New Guinea in relation to quarantine and disease diagnosis. There are legislative constraints, as the *Animal Disease Control Act* requires that imports of live fish come only from Australia and that only trout be imported. The current interest is in the farming of other species. In relation to quarantine requirements, Papua New Guinea does not have quarantine facilities or expertise for disease diagnosis and depends on outside institutions for this support.